Affect and Attention After Deleuze and Whitehead

New Perspectives in Ontology
Series Editors: Peter Gratton, Southeastern Louisiana University and
Sean J. McGrath, Memorial University of Newfoundland, Canada

Publishes the best new work on the question of being and the history of metaphysics

After the linguistic and structuralist turn of the twentieth century, a renaissance in metaphysics and ontology is occurring. Following in the wake of speculative realism and new materialism, this series aims to build on this renewed interest in perennial metaphysical questions, while opening up avenues of investigation long assumed to be closed. Working within the Continental tradition without being confined by it, the books in this series will move beyond the linguistic turn and re-think the oldest questions in a contemporary context. They will challenge old prejudices while drawing upon the speculative turn in post-Heideggerian ontology, the philosophy of nature and the philosophy of religion.

Editorial Advisory Board
Thomas J. J. Altizer, Maurizio Farraris, Paul Franks, Iain Hamilton Grant, Garth Green, Adrian Johnston, Catherine Malabou, Jeff Malpas, Marie-Eve Morin, Jeffrey Reid, Susan Ruddick, Michael Schulz, Hasana Sharp, Alison Stone, Peter Trawny, Uwe Voigt, Jason Wirth, Günter Zöller

Books available
The Political Theology of Schelling, Saitya Brata Das
Continental Realism and its Discontents, edited by Marie-Eve Morin
The Contingency of Necessity: Reason and God as Matters of Fact, Tyler Tritten
The Problem of Nature in Hegel's Final System, Wes Furlotte
Schelling's Naturalism: Motion, Space and the Volition of Thought, Ben Woodard
Thinking Nature: An Essay in Negative Ecology, Sean J. McGrath
Heidegger's Ontology of Events, James Bahoh
The Political Theology of Kierkegaard, Saitya Brata Das
The Schelling–Eschenmayer Controversy, 1801: Nature and Identity, Benjamin Berger and Daniel Whistler
Hölderlin's Philosophy of Nature, edited by Rochelle Tobias
The Late Schelling and the End of Christianity, Sean J. McGrath
Affect and Attention After Deleuze and Whitehead: Ecological Attunement, Russell J. Duvernoy

Books forthcoming
Schelling's Ontology of Powers, Charlotte Alderwick

www.edinburghuniversitypress.com/series/epnpio

Affect and Attention After Deleuze and Whitehead

Ecological Attunement

RUSSELL J. DUVERNOY

EDINBURGH
University Press

Edinburgh University Press is one of the leading university presses in the UK. We publish academic books and journals in our selected subject areas across the humanities and social sciences, combining cutting-edge scholarship with high editorial and production values to produce academic works of lasting importance. For more information visit our website: edinburghuniversitypress.com

Edinburgh University Press Ltd
The Tun – Holyrood Road, 12(2f) Jackson's Entry, Edinburgh EH8 8PJ

First published in hardback by Edinburgh University Press 2021

Typeset in 11/13 Adobe Garamond by
Servis Filmsetting Ltd, Stockport, Cheshire
and printed and bound by CPI Group (UK) Ltd,
Croydon, CR0 4YY

A CIP record for this book is available from the British Library

ISBN 978 1 4744 6691 2 (hardback)
ISBN 978 1 4744 6692 9 (paperback)
ISBN 978 1 4744 6694 3 (webready PDF)
ISBN 978 1 4744 6693 6 (epub)

Contents

For Ramona and Sebastian

Acknowledgements

It has been my great good fortune to be enriched by a wealth of philosophical interlocutors over the years that this book was in progress. Though my debts extend far further than can be fully acknowledged in a list of names, I want to extend special thanks to Ted Toadvine, who oversaw the dissertation that developed the first seeds of this book, and to Colin Koopman, Jason Wirth, Rika Dunlap, Elizabeth Sikes, David Craig, Phil Nelson, Bonnie Sheehey, Billy Goehring, Larry Busk, Scott Pratt, Naomi Zack, Beata Stawarska, Daniela Vallega-Neu, Justin Pack, Gus Skorburg, Rebekah Sinclair and Eric Severson for conversation, provocation and general companionship in the adventures of thinking. Thank you to Ahndraya Parlato for the cover photo. I would also like to thank all of my colleagues and teachers at the institutions I have been a part of during my formal philosophical practice: at the University of New Mexico (with a special thanks to Gino Signoracci, Paul Livingston and Russell Goodman), the University of Oregon and all my fellow faculty at Seattle University. Finally, I would like to acknowledge all of my teachers extending all the way back to elementary school whose tireless work is too rarely honoured.

The experience of working with the editorial team at Edinburgh University Press has also been a great pleasure. Special thanks to Carol Macdonald and Kirsty Woods for their guidance and expertise.

Finally, for their patience, good humour and support I am forever grateful to all of my extended family, and in particular my parents Russell and Nancy, my brothers and sisters-in-law, and my wife, Mollie, and children, Ramona and Sebastian, who remind me every day of what thought must never forget.

Abbreviations

For primary texts from Deleuze and Whitehead, I use the following abbreviations. (Note that the texts below are ordered by date of original publication in square brackets.)

DR   Deleuze, Gilles. 1994 [1968]. *Difference and Repetition*. Trans. Paul Patton. New York: Columbia University Press.

LS   Deleuze, Gilles. 1990 [1969]. *The Logic of Sense*. Trans. Mark Lester. New York: Columbia University Press.

PI   Deleuze, Gilles. 2001. *Pure Immanence: Essays on a Life*. Trans. Anne Boyman. New York; Cambridge, MA: Zone Books Distributed by The MIT Press.

DI   Deleuze, Gilles. 2004. *Desert Islands*. Ed. David Lapoujade. Trans. Michael Taormina. Los Angeles: Semiotext(e).

ATP   Deleuze, Gilles and Guattari, Félix. 1987 [1980]. *A Thousand Plateaus: Capitalism and Schizophrenia*. Trans. Brian Massumi. Minneapolis: University of Minnesota Press.

WIP   Deleuze, Gilles and Guattari, Félix. 1994 [1991]. *What Is Philosophy?* Trans. Hugh Tomlinson and Graham Burchell. New York: Columbia University Press.

CN   Whitehead, Alfred North. 1920. *The Concept of Nature*. Cambridge: Cambridge University Press.

SMW   Whitehead, Alfred North. 1967 [1925]. *Science and the Modern World*. New York: Free Press.

S   Whitehead, Alfred North. 1985 [1927]. *Symbolism: Its Meaning and Effect*. New York: Fordham University Press.

FR   Whitehead, Alfred North. 1958 [1929]. *The Function of Reason*. Boston, MA: Beacon Press.

PR   Whitehead, Alfred North. 1978 [1929]. *Process and Reality: An Essay in Cosmology.* Corrected ed. New York: Free Press.

AI   Whitehead, Alfred North. 1967 [1933]. *Adventures of Ideas.* New York: Free Press.

MT   Whitehead, Alfred North. 1966 [1938]. *Modes of Thought.* New York: Free Press.

Introduction:
Ecological Turbulence and the Adventure of Metaphysics

Attention, feeling and psychic ecology

This is a study of metaphysical resonances between two twentieth-century thinkers conducted in a spirit of speculative pragmatism. While it presumes an interest in its main figures (Alfred North Whitehead and Gilles Deleuze), it does not aspire to exhaustively detail all of their conceptual convergences and divergences, but rather to do something with the resonance it studies. As such, with a dose of caution and humility, it follows Deleuze's conception of philosophy as the creation of concepts. What concept comes out of a mutual reading of Deleuze and Whitehead? More specifically, how does taking seriously the process-inflected metaphysics implicated in their work potentially alter the lived experience of subjectivity?[1]

These questions are pursued along two main threads: (1) What does attention *do* in the context of a processual metaphysics? and (2) How does the metaphysical status of feeling and affect that comes out of Whitehead and Deleuze translate into existential orientation? Exploring these questions will take us into the thickets of their respective metaphysics, but the main implications can be summarised straightforwardly. With respect to attention, the thesis to be explored is that attention is ontologically creative, not just passively receptive. Furthermore, feeling and affect are ontologically prior to the consolidation of lived subjectivity, provided we understand that this claim does not map directly onto the customary category of emotion. The combined effects of these claims cut deeply against the grain of prevailing conceptual habits with regard to subjectivity. Taking them seriously, I argue, leads to an altered existential orientation which I describe as ecological attunement.

Because the resources of my argument are almost exclusively metaphysical, it remains speculative. It thus performs a sub-argument, presented in Chapter 1, about the importance of speculative philosophy. In a sense, the entire book can be read as an 'if . . . then' statement, with Part I corresponding to the 'if' and Part II corresponding to the 'then'. This is not to entirely jettison veridical questions, but it is to suggest that such questions are less interesting than a more open speculative exploration as offered here. How that interest inflects other lines of enquiry, especially empirical ones, is not pursued at present, though I hope that what I say might provoke those with better affordance to consider it. This is also not to say that the metaphysics of Whitehead and Deleuze are completely assimilated one into the other. While I do make the case for a significant family resemblance, there are undeniable differences. Because of my guiding questions, my approach to these differences is with regard to their consequences for existential orientation. So, while Chapter 2 for example identifies a tension in their respective characterisations of the 'transcendental' or 'virtual', my interest is how this difference impacts existential orientations.

To say that this book is resolutely speculative is not to pretend it occurs in a vacuum. On the contrary, it is acutely conscious of its unfolding amid intensifying global ecological crisis and accompanying social, political and existential turbulence. What does it mean to pursue speculative thinking in this context? This question is a broader version of one the book consistently returns to: how do concepts (in this case metaphysical concepts) inform our lives? How might different concepts lead to different ways of life? In this sense, while the book is not a direct engagement with ecological crisis or its accompanying social dysfunction and destabilisation, all of its speculative explorations are conducted with a long-term eye towards the necessity of deep transformation and change. Given present ecological conditions, significant change, whether cataclysmic or intentional, is unavoidable.[2]

Faced with the depth of such challenges, it is tempting to dismiss speculative thinking as an idle luxury better left for more tranquil times. However, while it would be foolish to claim that speculative thinking alone is sufficient, such thought has an important role in times of deep change and uncertainty. As Amitav Ghosh has observed:

> to imagine other forms of . . . existence is exactly the challenge that is posed by the climate crisis: for if there is any one thing that global warming has made perfectly clear it is that to think about the world as it is amounts to a formula for collective suicide. We need, rather, to envision what it might be. (2016: 128)

Avoiding a future along direst social, geopolitical and geophysical predictions will require multiple forms of social, political and material creation.

The stresses of these changes will leverage intense pressure on human psychic resilience. Will humans rise to their creative and spiritual potential or will these challenges induce lowest common denominator fear-based reactions?

Whitehead tells us that 'how the past perishes is how the future becomes' and that 'to perish is to assume a new function in the process of generation' (AI: 238, 291). The question facing us at present might be characterised as learning how to perish in a way that contributes to a different future, one that does not proceed according to the mystifications of global capitalism, hyper-consumption and logics of domination and exploitation. Might we attend to the ever-perishing of moments in a manner that learns how the very activity of attention may contribute to alternatives that evade what Isabelle Stengers aptly names 'barbarism' (2015: 50)? This claim does not pretend to be a magic bullet. While the book contributes to what Stengers describes as 'the art of paying attention' (2015: 100) in showing how our manner of attending and the objects we deem important are partially (not totally) a feature of metaphysical presumptions, it does not presume that this art alone is sufficient. Rather, it is a condition of transformation that must, ultimately, be connected with other forms of material and social change. This book is thus in one respect a propaedeutic to further investigation in the material and social. Nevertheless, this does not make it extraneous. Unless we learn how to attend differently, how to feel differently, how to, ultimately, develop different habits of subjectivity, we are unlikely to have the resilience necessary for transforming at the social and material levels. Uncertainty abounds, calling out for Deleuze's conception that 'to think is to experiment, but experimentation is always that which is in the process of coming about . . . What is in the process of coming about is no more what ends than begins' (WIP: 111, emphasis added). Speculative thinking is on the edge of this transition, constrained by the patterns out of which it emerges even as it seeks to alter them.

There is not one way to negotiate this transition, nor should there be. The very conception of a psychic ecology, in which patterns of subjectivity are emergent expressions of relational loops and social/affective/biophysical networks, mitigates against univocal normative prescriptions. But this is not to say that all ways of understanding subjectivity are equally valid. This tension between a principled openness to differing manifestations of ecological attunement while nevertheless not collapsing the concept into an underdetermined arbitrariness, remains in play throughout the entire text. It is, of course, isomorphic with worries about relativisms and dogmatisms that appear and reappear as a contemporary dialectic. Full treatment of this requires the entire book. At the outset it can be said that the study is guided by a constructive orientation. Rather than the final complete picture of reality, it sees metaphysical thinking as one form of revisioning and creating that is itself in progress and under construction. The principle of this openness is connection: how does this metaphysical study open ways of connecting with other emerging perspectives? In the context of crisis, this openness to connection is indispensable: 'There will be no response other than barbaric if we do not learn to couple together multiple divergent struggles and engagement in this process of creation, as hesitant and stammering as it may be' (Stengers 2015: 50).

A speculative concept: ecological attunement

Deleuze and Guattari tell us that: 'Wherever there are habits there are concepts' (WIP: 105). Habits are 'conventions' encoded in concepts and behaviour, which is to say that the creation of a concept is, in part, a proposal for new habits: 'To what convention is a given proposition due; what is the habit that constitutes its concept?' (WIP: 106). The first half of this book develops metaphysical reasons for the transformation of concepts of attention, feeling and affect. Taking these metaphysical results seriously has implications I characterise as an ecological attunement. A process metaphysical conception of subjectivity understands the psychic experiences constitutive of 'self' as inextricable from relational processes that are the ontological ground of all existence. Because the meaning and outcome of these processes are neither fixed nor predestined, activities of attention and feeling have an ontological role in the qualitative occurrences of occasions. Ecological attunement mobilises this transformed understanding of the ontological role of attention and feeling. Understanding attending as ontologically more basic than the abstract general category of self encourages modes of attention to qualitative micro-perceptions operative within ongoing events of experience.

To call such a mode of attuning ecological is to emphasise how it responds to two conceptual tensions inherent in thinking ecologically. The first of these has to do with one's relation to what is perceived as 'the whole' – whether this whole is characterised as Earth, the universe, or even Being as such. Environmentally inclined thinkers have sometimes told us that the key to a more sustainable way of life is greater attention given to one's relations to a whole that exceeds narrow self-interests. This certainly makes sense. Indeed, I would be inclined to say that it is correct, so far as it goes. But the manner of how we accomplish this is less decisive. Indeed, while we can, in principle, *conceive* of such an inclusive whole, this invariably and even constitutively requires reduction. We certainly cannot attune to the whole in any direct way because the very concept of the whole swallows up the dynamic differences that constitute our experience.

Consider a classic example from Deep Ecologist Arne Naess, who *begins* from a constituted self and argues for a move to extend the understanding of this self to recognise an *identification* with nonhuman others, from organisms and mountains to rivers: 'The ecosophical outlook is developed through an identification so deep that one's own self is no longer adequately delimited by the personal ego or the organism. One experiences oneself to be a genuine part of all life' (1989: 174). To be sure, such a claim is fully consonant with the spirit of an ecological attunement, but I will argue that the manner of orienting towards such an experience cannot be accomplished through a conceptual expansion of 'self' nor a desire to *identify* with a larger whole.

Instead, ecological attuning emphasises attention to the affective dimensions of experience that in a sense are prior to one's self, but not in the interests of identifying with a larger whole. This is a subtle distinction, but it is an important one if we are to avoid a facile identification with a whole modelled as inherently harmonious. Such a model, common to some strains of romantic environmentalism, is too disparate from living experience of conflict, discord and difference to hold up under scrutiny. As such, it enables less sympathetic interlocutors to dismiss a Naessian type call as merely wishful thinking.[3] Moreover, existential orientation towards a whole depends, to some extent, on how we *conceptualise* the whole insofar as such conceptualisation predisposes attention towards experience. Both Deleuze and Whitehead challenge static conceptions of part-whole relationships and replace them with dynamic and scale-relative processes accompanied by a metaphysical perspectivism (Chapters 2, 3 and 4).

This challenge also applies to how we conceptualise the self or subject. For ecological attuning to be a robust experiential potential, it cannot be characterised in terms of an expansion of self or interests. Instead, a more sustained scrutiny of the concept of self or subject must be conducted at a metaphysical level. Ecological attuning is not a unilateral expansion of agency or subjectivity but rather a revisioning of the metaphysical framework through which such concepts are understood.

The second tension for ecological attuning has to do with understanding the relationship between an ecology as fundamentally connected and in some sense unified, while also composed of a diversity of forms of life (which are forms of feeling). How is it possible to recognise the reality of a relational ecosystem without collapsing significant differences into a single rubric or measure? In Chapter 5, I show how this question points to an important issue of resonance between the biophysical and the social and economic, as manifest in what I diagnose (following Félix Guattari and Jean-Luc Nancy) as operations of equivalence. Where equivalence is oriented to produce a single rubric through which to quantify complexity, ecological attunement understands values to be produced in encounters of relations and primarily lived affectively.

A further manifestation of equivalence which ecological attunement must reconsider is the fixity of the subject/object polarity. Estonian bio-philosopher Jakob von Uexküll closes his influential *A Foray into the Worlds of Animals and Humans* with provocative remarks about this difficulty. Within an ecology, the same object plays different roles depending on relation and perspective. Uexküll describes an oak tree as at once resource for the forester, shelter for the fox, shelter, in a very different way, for the owl, source of shade, and so on. For the ant, the oak is a universe in itself, whereas for the forester, it is a one among many: 'the same parts [of the oak] are alternately large and small. Its wood is both hard and soft; it serves for attack and defense' (Uexküll 2010: 132). Though this point is intuitive, it remains difficult to translate into an existential orientation where the 'subject' typically occupies a privileged and fixed position. One does not like to think of one's self as an object, and, yet, in an ecology subject/object is a mobile distinction. As Uexküll asks: 'how does the subject exempt itself as an object in the different environments in which it plays an important role?' (2010: 126).

As is well known, Uexküll opens an expansive semiotics beyond the human by establishing heuristic categories for exploring the worlds of differently constructed 'subjects'. His categories mediate between images and tones – perception images, effect images, effect tones that are structured through different spaces (visual, haptic, olfactory) according to the subject's biological sense organs. However, Uexküll's framework is still Kantian, with categories naturalised by appeal to the sensory and biological organs. It is still the subject, whether human or not, that constitutes a world based on its filtering categories. Though Uexküll appeals to a musical metaphor for thinking the larger 'symphony' through which ecosystems interrelate, his approach has no real way of exploring the possibility of improvisations and themes transferring across the gaps between different forms of life. The symphony can only appeal to a pre-established harmony. Instead of a communicating ecology, forms of life are akin to one-note rule followers. By contrast, an ecological attunement bridges gaps induced by such a Kantian approach, since feeling forms or tertiary qualities mediate across life-form boundaries (Chapters 3, 4 and 6).

The tension between unity and diversity, between similarity and difference, is not eradicated by the approach taken here, but one's orientation to this tension shifts insofar as it becomes metaphysically constitutive. Increasing awareness of the multiple locations and scales a 'self' is enacted within helps to encounter the complexity of lived material and mental relations. This means, as developed in Chapter 2, that a subject as individuated is a result of a communication between at least two disparate fields and does not pre-exist this communication. If, as Uexküll suggests, human subjects tend to exempt themselves as objects, this is neither a necessary nor sound move, since we are participants in individuations that exceed us, and since we are composed, ourselves, of such participations.

Statement of purpose

This book is not an exhaustive study of the relationship between Whitehead and Deleuze. While it does both learn from and contribute to scholarly work on their intersection, this is always with an eye towards its impact on the concept under construction.[4] Unlike academic works where each chapter stands separately, the book makes a cumulative argument that amounts to a speculative proposal. While it is possible to read chapters selectively and they do stand on their own to some extent, the full effect of the argument requires reading the entirety.

In many respects, this book stands as a metaphysically motivated provocation towards existential rumination, using the work of Deleuze and Whitehead as its primary resources. In this sense, it is in keeping, I believe, with different aspects of the spirit of both. Undoubtedly both thinkers are characterised by conceptual audacity and willingness to experiment though they express this in diverging tones indicative of differences in sensibility and milieu. Their audacity is never in the service of absolutism or dogmatism, but rather what might be called a zest for thinking in relation with the problems of life. Both might be called 'existential metaphysicians' even as they transform the conventional connotation of the existential so that it is not limited to the human. As existential metaphysicians, the power of abstract thought is not in terms of whether it achieves the final answer, but rather insofar as it provokes, inspires or otherwise unsettles sedimented or stale habits to open us to the potentiality of creation. This book, in its own humble way, aspires to this conception as well.

Notes

1. I adopt the language of 'process' as a term of convenience noting that neither thinker explicitly identifies their thought in this way. Of course, Whitehead's identification with process metaphysics is inarguable and Deleuze's affinity for Whitehead is also increasingly recognised (Alliez 2004, Massumi 2014; Robinson 2009; Shaviro 2009, 2014; Stengers 2011, 2014; Williams 2005, 2009). Nevertheless, for reasons excellently

explored in Williams 2009, the goal is not to argue for an identification of Whitehead and Deleuze with a 'school' called 'process philosophy'.

2. Much change is of course already ongoing. The United Nations estimates 22.5 million people have been displaced by climate-related events since 2008 (United Nations Refugee Agency, 'Climate Change and Disasters' and 'Figures at a Glance'). Coastal communities worldwide face significant land loss, especially in the Far North. Extreme weather events (forest fires, droughts, heatwaves, hurricanes, flooding, polar freezes) have intensified in recent decades (Center for Climate and Energy Solutions, 'Extreme Weather and Climate Change' and Climate Communication Science and Outreach, 'Current Extreme Weather and Climate Change'). The 2018 IPCC (Intergovernmental Panel on Climate Change) report states that a 1.5°C raise in average temperature could occur as quickly as eleven years and certainly within twenty years without major reduction of global emissions. This poses significant risks to 'health, livelihoods, food security, water supply, [and] human security' (IPCC 2018).

3. My use of Naess as example is not intended as a rejection of his own nuanced investigation into environmental values. Like Whitehead, Naess understands the early modern distinction between primary, secondary and tertiary qualities to have led to confusion over the important of qualitative experience (1989: 51–2). Moreover, Naess insists that 'Confrontations between developers and conservers reveal difficulties in experiencing what is *real*. What a conservationist *sees* and experiences as *reality*, the developer typically does not see – and vice versa' (1989: 66). For this reason, Naess suggests that 'the philosophy of environmentalism [must] move from ethics to ontology' (1989: 67).

4. I frequently refer to this work in footnotes or the body of text. Of particular importance for the connection between Deleuze and Whitehead has been the work of James Williams, Didier Debaise and Isabelle Stengers.

Part I
Process Metaphysics and Ecological Attunement

Chapter 1

Motivating Metaphysics: From Radical Empiricism to Process

Risking the speculative: on problems and beginnings

Deleuze reminds us that beginnings in philosophy are perilous: 'Where to begin in philosophy has always – rightly – been regarded as a very delicate problem, for it means eliminating all presuppositions' (DR: 129). Most often for philosophy in the 'old style' (DR: xxi), such elimination is motivated epistemically. We seek to eliminate presuppositions when they propagate inherited illusions of *doxa*. Eliminating presuppositions is thus a prerequisite for gaining truth. Such methodological orientation, inflected through different contexts and aims, is constitutive for what it means to think philosophically: from Socrates to Descartes to Hegel and on. Philosophers question what is thought to be obvious or self-evident.

While such orientation is impeccable in principle, in practice its circularity is notorious. Deleuze accordingly distinguishes between 'objective presuppositions' and 'subjective' presuppositions. For the former, elimination through 'axiomatic rigor' is possible: this is the procedure of doctrinaire scientific method. But the results of such elimination still risk certifying presuppositions of the deeper lived paradigm within which their axioms are formulated. Getting to the level of such paradigmatic presuppositions also requires interrogating *subjective* presuppositions, which is far more challenging. *Can these be eliminated?* Given their entrenchment in experiential grain, the possibility of elimination requires first bringing them into view through questioning the obvious. Calling these presuppositions 'subjective and implicit' and 'contained in opinions rather than concepts' (DR: 129), Deleuze refers to the perennial efforts of the philosopher to bracket all lived attitudes, only for the Ourobouros-like elusiveness of this effort to return again and again. One philosopher's neutral begin-ning is exposed as full of presupposition by the next: as Hegel critiques Descartes, as Heidegger critiques Hegel, and the circle turns again.[1] For Deleuze, these criticisms have in common an 'attitude of refusing objec-tive presuppositions, but on condition of assuming just as many subjective presuppositions' (DR: 129).

Because what is deemed speculative depends on the audacity of its departure from consensus or convention, the status of presuppositions is all the more acute. This status cannot be settled in general; it requires understanding *what* is being challenged, *how* and *why*. There is no neutral ground upon which to easily demarcate the (pejoratively) speculative from the honestly respectable. Without jettisoning all sense of such demarca-tion, this emphasises that orientation towards the speculative is already entangled with pre-philosophical intuitions and complexities of a thinker's affective economy. Such pre-philosophical intuitions inform more devel-oped metaphysical inclinations, in particular with regard to one's sense of reality as, either, finished and determined, or, open and transform-ing (Parmenides or Heraclitus). These inclinations are not *determined* by affective economy, but a relation cannot be denied. The relevant affective tension is that between perceived safety or security and risk or danger. Undoubtedly, the speculative *is* risky, so the question is what risks are worth taking and why?

The stakes of presuppositions thus involve the very motivation of thinking. Why risk thinking? Can we think safely? To repeat, what risks (in thought) are worth taking and why? A dominant response to this question has been to seek the security of denying its relevance by estab-lishing apodictic certainty. There is no question of risk if thought discov-ers that which *cannot not* be true, the necessary, the eternal. The power of this ideal is undeniable, even if its outcomes since Kant tend to be presented in negative terms. That is, rather than positive articulation of the eternal, the philosopher establishes *necessary limits*. Only then, in the words of Kant, can metaphysics be set on the 'secure path of a science'.[2] Notwithstanding statistical procedures based on inductive probabilities nor the esoterica of contemporary physics' indeterminacies and relativi-ties, Kant's wording still reflects a popular image of science. This image privileges apodictic certainty as unquestioned good. In Whitehead's lan-guage, this ideal functions as a 'propositional lure' or 'lure for feeling' that motivates, attracts and organises intellectual endeavour.[3] While a certain moment of speculation might be necessary, its risk is underwritten by the promise of a knowledge defined through certainty. Such a knowledge minimises risk and maximises security in eliminating questions by discov-ering what-must-be-the-case.

This kind of response is impregnable in its internal logic. But for both Deleuze and Whitehead, it covers over what remains a decision to privilege a particular image of security as unassailable. Covering over this choice, that is, denying that it is a choice by pretending it is a universal good, *operates as a presupposition*. In particular, Deleuze and Whitehead are interested in the extent to which this tacit presupposition, in remaining invisible, blocks or inhibits creative alternatives. If the first response assumes the good of security as the ability to deny the relevance of the question 'what risks are worth taking and why', another choice is to *emphasise this question as always relevant*. The perpetual relevance of this question (what risks are worth taking and why) reflects the impossibility, in principle, of knowing what is ultimately possible ahead of time. It reflects the impossibility of a one size fits all answer to the question of what risks are worth taking and why. Instead of denying the relevance of this inherently unstable question, an alternative choice privileges speculative experimentation in the interests of creation precisely because of this instability. Certainty and creativity than appear in fundamental, mutually informing, tension. Certainty is aligned with safety, creativity is aligned with risk, and neither can fully extricate themselves from the other.

Negotiating this tension between mutually desirous goods (security and creativity) places the difficulty of presuppositions in a different light. It is no longer a question of *eliminating* presuppositions so much as it is one of studying them, testing them, experimenting with them. Such study and experimentation are no less rigorous than the presumed clarity of elimination. On the contrary, the goal of achieving the clarity of a first principle through the *elimination* of all presuppositions is easily conflated with a procedure of simplification in which we end up certifying what we already think. Indeed, in the hubris of claiming to eliminate presuppositions, we often entrench them more deeply and make their interrogation more unlikely. Speculative experimentation with presuppositions is a different manner of encountering the problem.

This encounter is not without its own observational, formal and affective challenges. Beginning with habits of observation, Whitehead reminds us that routine and assumption can render observation dull or impoverished. Because we 'observe by the method of difference' (PR: 4) we fail to notice that which does not present as departure from the ordinary while also presuming that we already understand the ordinary.[4] In this sense, the loop of observation is closed and repeats its own structuring assumptions. Whitehead's notion of philosophical speculation calls for more intense attention precisely because 'factors which are constantly present' are dismissed or not observed at all (PR: 5). His speculative concepts function as modes of 'imaginative thought' (PR: 5) designed to *heighten attention*.

In contrast to caricatures of idle fantasy or whimsical escape, Whitehead understands the need for heightened attention *under the influence of imaginative thought* as corrective to the omnipresent temptation to omit that which does not fit preconceived ideas. These preconceptions function at many levels, including prevailing formal or normative ideals. Whitehead is particularly concerned about the way in which the formal criterion of simplicity can tacitly structure attention. Because 'we are apt to fall into the error of thinking that the facts are simple because simplicity is the goal of our quest', Whitehead declares that 'the guiding motto in the life of every Natural philosopher should be, Seek simplicity and distrust it' (CN: 104).

The tendency towards simplification is reinforced by formal assumptions constitutive of what Deleuze calls the dominant 'image of thought'. The image of thought is structured around *the form* of representation as self-evident and obvious: '*Everybody knows, no one can deny*, is the form of representation and the discourse of the representative' (DR: 130). While it is true that 'the philosopher proceeds with greater disinterest' with regard to particular *content*, the *form* of representation still structures this variance by 'propos[ing] as universally recognized . . . what is meant by thinking, being, and self-in other words, not a particular this or that *but the form of representation and recognition in general*' (DR: 131, emphasis added). Despite its pervasiveness, Deleuze is keen to show that this form is neither necessary nor normatively neutral. Most importantly, the dominant image of thought presumes that 'thought has an affinity with the true; it formally possesses the true and materially wants the true' (DR: 131). This reinforces a 'moral' choice to privilege the 'same' at the expense of difference, thus bolstering the status quo and insulating thought against the risks of creation. Though it may require a fortitude of exertion, because thought's capacities are preordained to lead us to the true, these efforts remain secure and safe. The *cogitatio natura universalis* comes with a guarantee: if you play by the rules, you will achieve truth.

Assumptions of thought's 'good will' and 'upright nature' perpetuate an image of thinking as the operation of a faculty naturally inclined to find truth. Since thinking has this natural inclination, it can simply proceed in its ordinary modes. This preserves 'recognition' as thought's model: exercise your natural faculty to recognise what is already true. Importantly, this assumption prevails *at the formal level* and as such cannot be treated just by interrogating particular *content*. Indeed, Deleuze is not interested in wielding his diagnosis of the dogmatic image of thought as a means of rejecting specific philosophical claims. Such claims must be engaged and thought through on their own terms. Nevertheless, Deleuze can grant that the philosopher typically 'recognizes nothing in particular' while still arguing that recognition as model for thinking structurally inhibits ability to think beyond 'the recognizable and the recognized' (DR: 134).

By assimilating creativity and innovation onto familiar terms of already charted 'positions', thought does not confront the unknown and is not inspired to create beyond what has already been posed. This perpetuates a sense that the important philosophical problems are already understood: 'We are led to believe that problems are given ready-made' (DR: 158) and risks are confined to failing to solve a puzzle whose terms are not questioned. Such thinking 'recognizes only error as a possible misadventure . . . and reduces everything to the form of error' (DR: 148).[5] Denying thought's capacity to pose its own problems occludes its creative potential. But such creative potential is not inherent in the ordinary functioning of thought. As Deleuze claims: 'the problem is not to direct . . . a thought which pre-exists in principle . . . but to bring into being that which does not yet exist . . . To think is to create . . . but to create is first of all to engender "thinking" in thought' (DR: 147). This emphasis on the creative potential of thought is not neutral and carries its own risks of circularity. But in the context of presuppositions, it follows a coherent logic: assuming the terms of an inherited problem and accepting unquestioned the form of what counts as 'good' thinking is unlikely to force real thinking. Deleuze thus declares the dogmatic image of thought 'a hindrance to philosophy' (DR: 134). Indeed, because of the hegemony and invisibility of these formal assumptions in establishing obvious criteria, it is necessary that: 'someone – if only one – [has] . . . the necessary modesty [of] *not managing to know what everybody knows* ...' (DR: 130, emphasis added).

These observational and formal dimensions intersect in the affective. Confronting the heretofore unthought is dangerous, scary even. To 'not manage to know what everybody knows' invites vulnerability. Why submit to this risk? Indeed, the very endorsement of creative experimentation is already entangled with such affective diversity. Where one delights in mystery and the possibility of surprise, another shudders at a sense that everything we 'know' might be 'wrong'. Even as the unknown promises adventure, it is at once source of dread and anxiety.

In the context of such affective variability, Michèle Le Doeuff shows how Kant's thought is structured by an economy privileging certainty and safety in the name of secure 'possession' (1989: 8–20). Drawing on a psychoanalytic orientation, her analysis shows the extent to which values of certainty and safety function as unquestioned 'lures' that orient the excursions of Kant's thinking. While it would be foolish to reject an understandable desire for *a measure of safety*, this value is variable in its expression in a manner which Kant's characterisation of the risks of thought ignores.

Opening the third chapter of the Transcendental Analytic, Kant characterises his enquiry as having discovered an 'island' which is the 'land of truth' (Kant 1998 [1787]: 354/B294).[6] Such an island is the only safe terrain *for habitation*, whereas it is surrounded by 'a broad and stormy ocean' which is the 'true seat of illusion' (1998 [1787]: 354). Illusion is characterised by dynamism and inability to offer secure or certain harbour; instead, it is a domain 'where many a fog bank and rapidly melting iceberg pretend to be new lands' thus 'entwin[ing the voyager] in adventures from which he can never escape and yet also never bring to an end' (1998 [1787]: 354). Such a condition can only be a source of deception, and therefore the task of critical philosophy is to establish the limits of the understanding so as to 'distinguish whether certain questions lie within its horizon or not'. Without the certainty of these limits, we are never 'sure of its [reason's] claims and its *possession*' and 'must always reckon on many *embarrassing* corrections when it continually oversteps *the boundaries of its territory* (it is unavoidable) and loses itself in delusion and deceptions' (1998 [1787]: 355/B297, emphases added).[7]

There is a powerful lure to Kant's framing of the stakes of thinking, one that he makes explicit in stating that 'those who reject . . . the procedure of the critique of pure reason can have nothing else in mind except to throw off the fetters of science altogether, and to *transform work into play*, certainty into opinion, and philosophy into philodoxy' (1998 [1787]: 120, Bxxxvii, emphasis added).[8] But, without entering into the formidable details of his procedure, it is striking that this rhetorical device also displays a classic mode of prejudging common to the dogmatic image of thought – if you do not follow the 'one true' procedure, you can only be a charlatan. The point is not to convince those attracted to Kant's framing that it is wrong. Rather, I want only to highlight the extent to which this need not be a universal orientation towards the risks of speculative thinking. A privileging of possession perceived as immunisation against the 'embarrassment' of correction thus should not be presented as the sole viable starting point. This is why Whitehead characterises two basic orientations in philosophy in terms of their 'quarrel between safety and adventure' (MT: 173). Privileging safety leads to what Whitehead calls 'the fallacy of the perfect dictionary', in which it is held that 'mankind has consciously entertained all the fundamental ideas which are applicable to its experience' (MT: 173). This orientation follows the intuition of reality as determinate, fixed and essentially static. Whitehead instead endorses what he calls the 'speculative school', whose mission is to 'enlarge the dictionary' (MT: 173).[9] Rather than clarifying concepts that already exist (though this can be a necessary, but not sufficient, aspect of thought), the speculative school rejects the presupposition that we already have all of the concepts needed to understand reality. Implicitly, the speculative attitude follows an intuition of the real as dynamic and in process rather than static or fixed. If this is the case, certainty becomes more suspicious since it likely involves affirming static and partial propositions as falsely comprehensive.

Just as Deleuze worries that the dogmatic image of thought hinders creative alternatives, Whitehead's choice of 'adventure' is motivated by a sense that *the risk is worth it* insofar as we cannot presume that we have achieved certainty about the nature of reality: 'the chief error in philosophy is overstatement . . . [where] the estimate of success is exaggerated' (PR: 7). Creative alternatives, for both Deleuze and Whitehead, are motivated by a spirit of metaphysics as seeking to approach the real independent of inherited convention or orthodoxy. Whitehead indeed does not hesitate to refer to speculative philosophy as 'seeking the essence of the universe' (PR: 4). But we cannot presume that this essence is easily accessed through language or proceeds from traditional first principles: 'Philosophers can never hope finally to formulate . . . metaphysical first principles' because 'deficiencies of language stand in the way inexorably' (PR: 4).

Deleuze and Whitehead remain vulnerable to sceptical attacks by those wedded to terms and logics they are trying to innovate beyond because they understand that philosophical thought impervious to objection is unlikely *to do* anything at all. When Kant deems the speculative as 'the childish endeavor of chasing after soap bubbles' (Kant 1997 [1783]: 44), he presumes the only criterion for success is certainty in the name of possession. Whitehead and Deleuze reject this. In a clear case of Whiteheadean influence, Deleuze writes, 'Philosophy does not consist in knowing and is not inspired by truth. Rather, it is categories like Interesting, Remarkable, or Important that determine success or failure' (WIP: 82).[10]

Rejecting universal first principles is defiantly not a rejection of metaphysics as such. The motivation for 'not managing to know what everybody knows' is thus not primarily negative or deconstructive. Rather, it is in the context of efforts to think the real while acknowledging that ordinary habits of thought (what everybody knows) hinder this effort. The test for creation is thus not merely novelty for novelty's sake. Rather, as Whitehead writes, 'the speculative school appeals to direct insight, and endeavors to indicate its meanings by further appeal to situations which promote such specific insights' (MT: 173).

Everything depends on how we understand the status of such appeal to direct insight. Given Whitehead's discussions of the challenging intersection between language and speculative thought, it is too hasty to understand such direct insight as translating unequivocally into its propositional representations. If it did, Whitehead would be susceptible to Sellars's critique of the myth of the given or charges of naïve romanticism or mysticism. When such 'direct insight' is appealed to as ineffable evidence, then thinking ends rather than begins. For Whitehead, such appeals do not serve as conceptual *justifications*, but rather as invitations to attention. The insight in question is not the conclusion of a discursive argument, but rather the experiential presence out of which discursive thought and argumentation emerge.

Concepts might open a more rigorous attention to this experiential presence, or they might reduce, close or deny it. For Deleuze and Whitehead, if philosophy is to do more than repeat sterile presuppositions, it must acknowledge there is more going on than recognition of something already known. Deleuze writes, 'Something in the world forces us to think. This something is an object not of recognition but of a fundamental encounter' (DR: 139). An encounter is not governed by categorisation into pre-existent categories, but rather exceeds conceptual categories in being fundamentally affective *before* it is representational. Such an encounter 'may be grasped in a range of affective tones: wonder, love, hatred, suffering. In whichever tone, its primary characteristic is that it can only be sensed' (DR: 139). Depictions of experience which discount the way affective encounters serve as pre-representative ground are just skimming the surface.

Entrenched habits of thought enable this superficiality by insulating the thinker from risks inherent in encountering what exceeds the known or understood. This is why Whitehead lists 'trust in language as an adequate expression of propositions' along with 'the subject-predicate form of expression' as two 'habits of thought' that his speculative philosophy is most concerned to challenge (PR: xiii). Such habits, so embedded as to be almost invisible, are paradigmatic of the form of 'what everybody knows'. This is notably demonstrated in the way that the 'overemphasis on Aristotle's logic' repeatedly projects a subject-predicate logical and grammatical form onto metaphysical thinking, a worry that is an operative premise for all of Whitehead's work.[11] This manifests in the metaphysical habit of 'the ingrained tendency to postulate a substratum for whatever is disclosed in sense-awareness' (CN: 12) and the way that grammar structures casual logic. Most specifically, the ontological prioritising of nouns as the real substances to which contingent verbs occur is so entrenched as to appear self-evident. This conflates a 'convenient form of speech' with a subject-predicate form deemed metaphysically universal. Whitehead observes:

> Predication is a muddled notion confusing many different relations under a convenient form of speech . . . the relation of green to a blade of grass is entirely different from the relation of green to the event which is the life history of that blade for some short period and is different from the relation of the blade to that event. (CN: 12)

The radicality of this observation highlights the tension between 'managing to not know what everybody knows' and expressing the efforts of this thought through received linguistic tools. Despite its deficiencies, language remains necessary for the expression of thinking even as the speculative philosopher challenges presuppositions encouraged in its grammar. The speculative philosopher employs language while remaining on guard against its tendencies towards reification, careless generalisation and slippage between grammar and metaphysics. As Whitehead says, 'words and phrases must be stretched towards a generality foreign to their ordinary usage; and however such elements of language be stabilized as technicalities, they remain metaphors mutely appealing for an imaginative leap' (PR: 4).

Imaginative leaps are notoriously tricky. They occur before the backdrop of what Isabelle Stengers refers to as 'dual temptations': while they clearly resist construing philosophy in 'the [Kantian] role of guardian of rationality', they also must evade the temptation of purporting to 'escap[e] rationality through the pathos of inspiration or emotion' (2014: 188). Thus, when Whitehead insists that 'the aim of philosophy is sheer disclosure' (MT: 49) and the 'eliciting of self-evidence' (MT: 107), such disclosures thread a gap between conventional presuppositions (what everybody knows) *and* the interior experience of the inspired. It may be the case that this interior inspiration is an effect of Whiteheadean or Deleuzean ideas, but the work of the speculative philosopher must be to concretise such inspiration through skilful use of concepts. If the imaginative leap cannot be avoided, just leaping alone is not the goal.

Whitehead's imaginative leap is thus leavened by a need for observation and a principled fallibilism: 'The true method of discovery is like the flight of an aeroplane. It starts from the ground of particular observation, it makes flight in the thin air of imaginative generalization; and it lands again for renewed observation rendered acute by rational interpretation' (PR: 5). Without these constraints, experimental iconoclasm can be over-celebrated uncritically as end-in-itself. Such celebration of the novel for the novel's sake is easily co-opted and encouraged by the procedures of IWC (Integrated World Capitalism)[12] and its manufactured thirst for the new trend, the new style, the new fad. Without care, adoration of the new can be just as philosophically vacuous and stultifying as the dogmatic image of thought.[13] Leap-taking effects engendered by speculative constructions invite careful consideration of their manners of responding to problems. Challenging 'what everybody knows' requires locating a problem in which what everybody knows is, at least in part, constitutive of the problem.

Attending to the bifurcation of nature

At the outset of *The Concept of Nature*, Whitehead names the 'bifurcation of nature' as the 'fallacy' with which 'modern natural philosophy is shot through and through' (CN: ix).[14] Rather than static errors in content alone, fallacies are habits of thought that effect what it is possible to think. Fallacies *do something* to thought by structuring the framings of abiding problems.

The *Concept of Nature* is a strange and beautiful book that has taken on a renewed life in the context of recent Whitehead interest inflected through Isabelle Stengers and Deleuze.[15] The book's impetus is the inaugural course of Tarner Lectures at Trinity College given in the fall of 1919, whose topic is stated as: 'the philosophy of the Sciences and the Relations or Want of Relations between the different Departments of Knowledge' (CN: 1). The text that results, published in 1920, forms, as he puts it, a 'companion volume' to his 1919 *An Enquiry Concerning the Principles of Natural Knowledge*. Both books display Whitehead's training as a mathematician and his interest in emerging relativity theories in physics. They are written, in some respects, before Whitehead becomes a 'philosopher', or at least before this vocation becomes officially recognised in the invitation to join the Harvard philosophy department.

Be that as it may, the book sparkles with kernels of innovation and insight that become more technically developed in Whitehead's later philosophy.[16] Whitehead states that his object 'is to lay the basis of a natural philosophy which is the necessary presupposition of a reorganized speculative physics' (CN: x). But, he also takes pains to stress that he is not (yet) doing metaphysics but rather 'endeavoring in these lectures to limit ourselves to nature itself and not to travel beyond entities which are disclosed in sense-awareness' (CN: 19).

Given the topics of the discussion (space, time, and so on) it can be difficult to see how Whitehead can maintain he is not doing metaphysics, a difficulty he recognises: 'it is difficult for a philosopher to realise that anyone really is confining his discussion within the limits that I have set before you. The boundary is set up just where he is beginning to get excited' (CN: 32). What are these limits and what do they tell us about the implications of the problem Whitehead is constructing? His topic is nature, defined as 'that which we observe in perception through the senses … [in which] we are aware of something which is not thought and which is self-contained for thought' (CN: 2). The limit involves not asking about the perceiver or process of perception, but only the perceived (CN: 20). What is really perceived? This would appear to be a kind of phenomenology. Nevertheless, its implications go beyond phenomenology as well as appearing to go beyond the claimed limits of the text.

Whitehead's fundamental claim is that 'the concrete elements of nature [are] events' (CN: 22) such that, even time and space are 'abstractions' from these concrete elements. This is to say that what is perceived, what is before us in perception, are events. As such, this is a radical empiricist claim. But Whitehead is concerned to think through, already, what this can mean for ordinary habits of thinking. He sets up two constraints that the speculative constructions of the later texts (eternal objects and actual occasions) endeavour to meet. Events as the concrete elements of nature require two modes of transition. One we might call global – the 'passing of nature, its development, its creative advance' and the second local – 'the extensive relation between events' (CN: 23). It is these two 'facts' and their relation that must be explained in a coherent philosophy of nature. To use a physical analogy – the train passes along the track, within the train, passengers participate in overlapping, but not equivalent, events of personal perception. Both events are entangled in each other.

This looks a lot like metaphysics. Whitehead's refusal of that term is an effort to create a hesitation so that what he is endeavouring to bring into view is not immediately enlisted into categories already established (chiefly matter and mind as oppositional categories). Indeed, he is at once subtly criticising these terms but not (yet) offering a replacement. This is where the bifurcation of nature becomes so important, precisely because it operates according to the self-evidence of categories that Whitehead believes function against a real reckoning with the events of sense perception and nature.

Given the topic of the lecture series and Whitehead's professional context, the bifurcation can be read as primarily an epistemic difficulty for a unified philosophy of the physical sciences. This framing is ultimately insufficient for emphasising its most pertinent existential and experiential effects, but it can be used as an entrance. As epistemic difficulty, the bifurcation involves an inconsistency within the paradigm of the natural sciences taken to be 'concerned with the adventures of material entities in space and time' (CN: 11).[17] Defining the natural sciences in this way presumes the cogency of a concept of matter that is already separate from phenomenal experience of it. Since the natural sciences endorse empirical observation as privileged means of verification, there is at least a constitutive tension, if not outright incoherence: why would an empiricist orientation, which purportedly grounds all ideas in sensational experience, generate an ontology that undercuts the reality of nearly all of the qualitative features of this experience? As Whitehead asks, 'Why should we perceive secondary qualities? It seems an extremely unfortunate arrangement that we should perceive a lot of things that are not there. Yet this is what the theory of secondary qualities in fact comes to' (CN: 27).

Whitehead summarises the problem:

> What I am essentially protesting against is the bifurcation of nature into two systems of reality, which, in so far as they are real, are real in different senses. One reality would be the entities such as electrons which are the study of speculative physics. This would be the reality which is there for knowledge, although on this theory it is never known. For what is known is the other sort of reality, which is the byplay of mind. Thus there would be two natures, one is the conjecture and the other is the dream. (CN: 21)[18]

An account of nature cannot begin from a presumption of a categorical (and unexplained) difference between disparate events of perception that provide access to presumed material entities of nature: 'there is but one nature, namely the nature that is before us in perceptual knowledge' (CN: 27). This is not to say that perceptual 'knowledge' *exhausts* nature, but it does establish a constraint: events of sense perception must themselves be understood as consistent with what will be called nature. To fail this constraint would be to install a disjunction that sets up oppositions (between mind and matter, for example) *as given* when they are in fact abstracted from a more inclusive experience: 'we may not pick and choose . . . the red glow of the sunset should be as much part of nature as are the molecules and electric waves by which men of science would explain the phenomenon' (CN: 20).

Given Whitehead's language of 'two systems of reality', the bifurcation might also be read as a metaphysical dualism. Yet, understanding the bifurcation in this way misses the experiential depth of the diagnosis and makes Whitehead's later metaphysical innovations appear more arbitrary than they are. As Didier Debaise argues, understanding the bifurcation as equivalent to dualism in effect mistakes an effect for a cause. Dualism names a result produced by the operation of bifurcation. This operation takes what is experienced in a mode of togetherness and bifurcates it through abstractive analysis. The bifurcation occurs *in* experience and produces *experiential* effects (Debaise 2017b: 24–5). Conflating this operation with a metaphysical position (dualism) reinforces the idea of an external vantage, as if the thinker is selecting from possible puzzle solutions, when instead what is at stake is the manner in which bifurcation structures living experience. Indeed, bifurcation, as mode of abstraction, is so entrenched it is no longer perceived as operation. Its experiential effects are rendered *given* rather than produced, displaying the power of bifurcation to produce effects that reinforce its operation.

Bifurcation is a fallacy when it forgets that what it splits was experienced together: 'The nature which is in fact apprehended in awareness holds within it the greenness of the trees, the song of the birds, the warmth of the sun, the hardness of the chairs, the feel of the velvet' (CN: 21). Paradoxically, bifurcation of nature generates a division between 'conjectured' reasons for qualities of experience that are taken as more real: 'The nature which is the cause of awareness is the conjectured system of molecules and electrons which so affects the mind as to produce the awareness of apparent nature.' When reified and read back into experience, greenness and warmth are relegated to lesser status of the merely 'subjective'. A conjecture of reason (molecules and electrons) is privileged and felt experience is downgraded to the epiphenomenal. At certain theoretical extremes, these felt aspects are dismissed in the name of 'hard' science. The potential for a reinforcing feedback loop is enormous: if certain qualities are deemed less real, there is a tacit implication that they are less important to attend to. This can alter attention *in* experience.

The bifurcation is a paradigmatic instance of the power of abstraction to produce lived effects. This is not categorically bad. In many respects what Whitehead's philosophy offers is a pragmatics of abstraction. Debaise writes 'inquiry into the mode of existence of abstractions and their function in the most concrete experiences . . . is fundamental to the philosophy of Whitehead' (2017b: 23). In the case of the bifurcation, the effects of the abstraction fail on both epistemic and existential grounds. Indeed, the bifurcation, for reasons sketched above, is doomed to *necessarily* reduce experience because it enables a dismissal of attention without providing a coherent reason for it. It is a presupposition about experience that orders how we prioritise observation that cannot itself be explained in observation. Worse, it is in direct tension with the felt intuition that: 'by due attention, more can be found in nature than that which is observed at first sight' (CN: 20). By prioritising certain aspects of observation and dismissing others, its effects are not only manifest in relations to phenomenal or qualitative experience, but even in how we conceptualise ourselves. *Impoverishing the perceived potentials of attention, these effects alienate the subject from implications of their lived experience, especially at intuitive, felt, and embodied levels.*

Diagnosing bifurcation as operation of abstraction might suggest seeking its resolution in the non-abstract, a call for a return to concrete experience as it really is, unmediated by abstraction. This is emphatically not Whitehead's strategy. Such a call does not understand the constructive nature of experience. Abstraction, in some form, is a constitutive operation of experience. The resolution cannot be to erase it as such, but rather become more critical about its activity. Whitehead observes,

'You cannot think without abstractions; accordingly, it is of the utmost importance to be vigilant in critically revising your mode of abstraction' (SMW: 59).[19] The very necessity of abstractions makes understanding their function imperative. Abstractions are not 'things' that lay over experience, they are activities that occur as partially constitutive modes of experience that shape it: 'abstractions are neither representations nor generalizations of empirical states of affairs but constructions' (Debaise 2017b: 23).

Despite Whitehead's affinity with the many twentieth-century critiques of metaphysical dualism (whether eco-feminist (Val Plumwood and others) Deweyan, Bergsonian, or phenomenological (Merleau-Ponty)), what is required is not just criticism but deeper study of its production so as to *construct* a different way out of this problem. Critiques of dualism as such tempt a mere switching of priority of the poles or assimilation of one pole into the other (hence the idealism-materialism dialectic that post-Cartesian, post-Kantian philosophy remains captured by). But the need for an alternative construction is not readily granted. Die-hard proponents of the bifurcation may be too satisfied with its power to see need for an alternative abstraction. If you don't already value the greenness of the foliage then why do you care about its loss? Are we not mistaking an axiological question for a physical one? Isn't it indeed the case that the greenness of the leaves is not in the leaves at all? And yet the (presumed vicious) circularity cuts both ways. If the critic of bifurcation is charged with question-begging in presuming greenness is valuable, the enthusiastic bifurcator can also be charged with question-begging in presuming what is at stake is locating the substance bearer of the property of green-ness. *Both sides in this sense are bifurcating.* Evading the bifurcation cannot mean simply affirming the qualitative as epistemic end-in-itself, but nor can it be just accepting that extended material objects index mind-independent reality as such. Both are putting the cart before the horse by presuming that the problem, as commonly constructed, is already decisive.

In *The Concept of Nature*, Whitehead develops the possibility of an alternative abstraction for thinking about perception and reality: events.[20] This conceptual shift performs a different operation of abstraction. The goal is an approach to experience that is rigorous, inclusive and attentive while avoiding the incoherence perpetuated by bifurcation. Because of the extent to which the bifurcation is taken for granted, the positing of an alternative abstraction must appear, at least at first, as speculative. However, while it is true that ultimately Whitehead goes much further in unfolding implications of this conceptual shift, its impetus remains *experiential* and can be situated in a lineage of radical empiricism.

Pure experience as series and events

A distinguishing characteristic of attention to the Deleuze/Whitehead intersection is its emphasis on alternative empiricisms.[21] If the bifurcation shows how operations of abstraction can hinder, block or prefigure experiential attention, can we, to creatively paraphrase Samuel Beckett, 'abstract again, abstract better'.[22] Calling such a proposal empiricist names a commitment to experience as source and constraint for abstraction. Alternative abstraction must find its reasons for being in experience even as metaphysical innovations developed through such abstractions challenge the conventional standing of distinctions (between inner and outer or between self and world) inherent in ordinary concepts of experience.

I will explore this complementarity by focusing on William James's concept of 'pure experience'.[23] James's radical empiricism includes both an inclusive methodological orientation *and* a positive speculative proposal exemplified in the concept of pure experience. While the former is widely noted, it is the latter that is most important for understanding the linkage between James's radical empiricism and the metaphysical turns to events developed in Whitehead and Deleuze.[24]

The methodological orientation of radical empiricism is most celebrated for its affirmation of continuities in experience. Following the postulate that 'everything real must be experience-able somewhere, and every kind of thing experienced must somewhere be real' (James 1976: 81), James challenges the atomistic emphasis of the British Associationist tradition since Hume.[25] This tradition displays a phenomenological error common to 'ordinary empiricism' in: 'do[ing] away with the connections of things, and . . . insist[ing] most on disjunctions' (James 1976: 22–3). James insists instead that '. . . relations between things, conjunctive as well as disjunctive, *are just as much matters of direct particular experience*' (1976: 7, emphasis added).[26] This demonstrates a commitment to inclusivity: there can be no in-principle dismissal of any aspect of experience (the vague, dull, obscure, felt or intensive, even the hallucinatory) as *a priori* less than real. Whitehead exhibits this radical empiricist orientation in exclaiming:

> we must appeal to evidence relating to every variety of occasion. Nothing can be omitted, experience drunk and experience sober, experience sleeping and experience waking, experience drowsy and experience wide-awake, experience self-conscious and experience self-forgetful, experience intellectual and experience physical, experience religious and experience skeptical, experience anxious and experience care-free, experience anticipatory and experience retrospective, experience happy and experience grieving,

experience dominated by emotion and experience under self-restraint, experience in the light and experience in the dark, experience normal and experience abnormal. (AI: 226)

Such a methodological orientation alone is not however sufficient for
understanding the full radicality of the concept 'pure experience'. Though
in part an invitation to perspicuous attention to experience in all its variability, confusion and ambiguity, its positive implications go well beyond
such a call. Pure experience indeed names 'the immediate flux of life
which furnishes the material to our later reflection with its conceptual
categories' (James 1976: 46), but such immediacy does not certify determinative normative or epistemological purchase. The problem generating
James's proposal of pure experience is not *simply* a false description but the
pervasiveness of 'vicious intellectualism' in structuring the lens through
which the descriptive emerges.[27] The proposal of pure experience thus
indicates the complexity of intersection *between* the immediate flux and
the concepts through which it is described. It is not that certain concepts
are 'false' and pure experience names the true, rather, pure experience is a
gesture towards renewed attention to the real that illuminates a conceptual
problem hidden in ordinary modes of description.

Specifically, James is inspecting the conceptual polarity between subject and object. In pure experience, he stresses 'there is no self-splitting
into consciousness and what consciousness is "of." Its subjectivity and
objectivity are functional attributes solely, realized only when the experience is "taken," i.e., talked-of' (James 1976: 13). While James emphasises
how pure experience shifts conceptualisations of consciousness, the shift
is not exclusively internal to subjectivity or consciousness. If it were, then
pure experience would be a strategic device for phenomenological purposes, continuous with James's earlier work in *Principles of Psychology*.
Pure experience might then get us to look more closely at internal consciousness, but entirely within the framework of the subject/object distinction as fundamental.

James goes much further than this in *Essays in Radical Empiricism*. Pure
experience is explicitly presented in *speculative metaphysical terms* in which
'we start with the supposition that there is only one primal stuff . . . in
the world . . . of which everything is composed, and . . . we call that stuff
"pure experience,"' (James 1976: 4). With this supposition, polarities of
subjectivity and objectivity are *not* naming fundamental ontological substances or categories, but rather functions of relations between different
moments of pure experience. Consciousness does not name 'an aboriginal
stuff or quality of being . . . out of which are thoughts are made', rather,
it is a name for a 'function in experience' (James 1976: 4). Rather than

a *res cogito*, the mental as substantially different in kind, pure experience implies a shift to relational events as ontologically primary. Knowing is a kind of event, 'a particular sort of relation . . . into which portions of pure experience may enter' (James 1976: 4).

Given James's emphasis on consciousness and the knowing-relation, it is often missed that he extends this suggestion to *both* 'minds' and 'objects'. Just as mental events express particular sorts of relations within the broader category of pure experience (which, recall, is inclusive of all reality), *material* objects are also 'processes' of relations within this category. This is apparent in his response to the problem of doubling posed by standard accounts of mental representation. Perception appears to produce a doubling: there is one physical object and one's perception of it as a mental representation. This is the basic 'dualism' of experience, resulting in 'the paradox that what is evidently one reality should be in two places at once both in outer space and in a person's mind' (James 1976: 8).

James proposes that pure experience 'reinterprets' this dualism – there is still a dualism, but it no longer revolves around a fundamental ontological gap between two different substances. Rather, it is what might be called a local or situational dualism in which the 'same' 'bit of experience' can play dual roles depending on context (James 1976: 8). Instead of one 'mysterious and elusive' fundamental gap, there are now multiple gaps of relations or perspectives within one inclusive metaphysical category of pure experience. The bits of pure experience are events or processes, not self-standing things:

> The puzzle of how the one identical room can be in two places is at bottom just the puzzle of how one identical point can be on two lines. It can, if it be situated at their intersection; and similarly, if the 'pure experience' of the room were a place of intersection of two processes, which connected it with different groups of associates respectively . . . (James 1976: 8)

The 'reader's personal biography' as one series of events intersects with another series of events ('carpentering, papering, furnishing, etc.') (James 1976: 9–10).

This shifting in the terms of the problem leads Whitehead to compare James's 1904 'Does Consciousness Exist?' to Descartes's *Discourse on Method* in announcing 'the inauguration of a new stage in philosophy' that 'clear[s] the stage of the old paraphernalia' (SMW: 143). Rather than explaining a relation between two categorically different substances, we now have to explain relevant distinctions between different kinds of event relations. The first problem follows a logic of exclusive disjunction and determinate negation, the second may follow a logic of degrees. Some relations will be more 'mental' than others, some will be more 'physical' than others, but we do not presume that these descriptions mark all-or-nothing sets of opposing substances. Instead, they mark tendencies, degrees, as James puts it: 'relations are of different degrees of intimacy' (1976: 23).

This shift is still not widely understood. As Whitehead says, if James 'open[s] an epoch by clear formulation of terms in which thought could profitably express itself' (SMW: 147), the profits of that thought are still in the making.[28] Most radically, stressing the speculative posit of pure experience allows for an empiricist orientation without the necessity of a categorical subject. Challenging the idea that experience is necessarily a predicate of a pre-existent subject, this speculative posit flies in the face of convention. One task of the present text is showing why it is worth the risks to explore such a paradoxical notion as asubjective experience. This requires more sustained investigation into metaphysics without presuming the indubitability of the subjective/objective pivot as necessary starting point.

This alternative to beginning from categorical opposition and determinate negation attracts Jean Wahl, a notable early French reader of James and Whitehead and influence on Deleuze.[29] A consistent feature of Wahl's reading is his interest in how a radical empiricism keeps the concrete immediacy of experience as vital and open while not necessarily falling prey to the Hegelian critique of the immediacy of the 'here' and 'now'.[30] As such, Wahl aligns himself with 'realism' rather than the quasi-Hegelian idealisms prominent in his milieu, but this realism is understood as closely aligned with pluralism. This is first demonstrated in how Wahl's 1920 *Les Philosophies pluralists d'Angleterre et d'Amérique* links the implications of radical empiricism to a pluralistic realism.[31] Wahl consistently emphasises the connection between radical empiricism and metaphysics as the most important aspect of James thought: '"pragmatism" regarded as the consideration of precise consequences, is but the logical result of that whereof pluralism is the metaphysical result' (Wahl 1925: 117).[32]

Wahl's interest in the way that James (and by extension Whitehead) provides a different approach to experience is further demonstrated in his 1932 study *Vers le Concret*.[33] Here, Wahl argues that James and Whitehead offer a genuine third way that resists the recoil between idealism and realism structuring the thought of his milieu.[34] This recoil is oriented by a Hegelian logic of determinate negation wielded in a dialectic between thought and world with an ultimate goal of arrival, through thought, at the absolute. Wahl wants a realism that is not reductive or easily captured by the Hegelian critique of 'immediacy'. Pure experience attracts him insofar as it enables such a realism. The key is that attention to pure experience reveals *events* as a 'mixture of the continuous and the discontinuous that defines a rhythm, a volume, a person' (Wahl 2017: 37). This is 'precisely what [Whitehead, James] grasp . . . blocks of duration, volumes, *events*' (Wahl 2017: 37, emphasis added). If 'these philosophers [James and Whitehead] claim the rights of the immediate', it is an immediacy of events that come in different degrees of intensity and intimacy. Rather than the here and now taken as complete in itself, there is a 'density' in the concrete characterised as a mixture of tendencies (the continuous and the discontinuous enveloped in one another) (Wahl 2017: 48–9). It is this tension, between a desire to maintain contact with an extra-conceptual real and a recognition that it inherently slips from grasp as soon as we say anything about it – that attracts Wahl to James, and that sets what Wahl will still call a 'dialectic' in a different direction. As Wahl writes, 'The concrete will never be something given to the philosopher . . . [but] it will be what is being pursued' (2017: 51).

Pure experience thus brings into view a different kind of dynamic realism, one whose pluralism challenges stable identity fixed independently of perspective that is a feature of many 'common sense' realisms. This follows the above citation regarding how one 'identical' room can be in two places when conceptualised as relational event at the intersection of two (or more) different series of events. In the context of evading a single categorical gap between mind and world, the event posit includes both minds and world as abstractions constructed out of processes and events. This raises the possibility of an excess to any particular relation-event-experience, since, as James notes, it is 'a member of diverse processes that can be followed away from it along entirely different lines' (James 1976: 8). There is the ongoing room event in the material series of the history of the wood, not to mention the many room events all occurring along the series of the personal histories of its occupants. And these continue to happen, in principle. Deleuze, for example, alludes to James in declaring that '... perspectivism amounts to a relativism, but not the relativism we take for granted. It is not a variation of truth according to the subject, but the condition in which the truth of a variation appears to the subject' (Deleuze 1993: 20). When experience is the juncture of lines of functions any 'point' (a point of view in a particular place at a particular time) is an intersection of many different lines (or series, in Deleuze's language).

Rather than subjective relativism, James's radical empiricism implies a perspectival realism. The relativity in question is not only relative to one fixed pole (the subject) but is rather multiplied into relativities at any number of potential loci of experience. This will be more rigorously developed in Whitehead, but already in James there is a multiplication of relational loci as different 'elements' of pure experience. This perspectival relationality complicates determinate negation's operation on two-term relations. Events are instead relationally promiscuous. This promiscuity remains a problem for formal logical syntax (of which Whitehead is well aware), because, with some exceptions, event relations can only be reductively described in two terms. But the radical empiricist orientation affirms this multiplicity as intrinsic to experience. The burden of multiplicity's challenge lies on the side of formal syntax, not a denial of the reality or relevance of this promiscuity.

As Wahl recognises, the multiplicity of experience indicates a paradox: a concrete experience is at once irreducibly precise and yet cannot be characterised with discrete precision. What Wahl proposes, and what he sees in James and Whitehead and passes on to Deleuze, is an image of a dialectic of thought that is not just unfolding conceptual oppositions, but that is always attempting to move in relation to something that is not thought. The dynamism of thinking is not internal alone but is rather an expression of a processual reality: 'Movement is not immanent to the idea [alone] . . . [rather] it comes from what the idea tries to do *in relation to something other than itself*' (Wahl 2017: 51, emphasis added). Deleuze characterises this as Wahl's commitment to difference: 'All of Jean Wahl's work is a profound meditation on difference . . . [and] the irreducibility of the different to the simple negative; on the *non-Hegelian* relations between affirmation and negation' (DR: 311, n18). The influence of Wahl's reading of James on Deleuze cannot be overstated. Indeed, Deleuze is entirely Wahlian in declaring that radical empiricism '*presents only events*' (WIP: 47–8, emphasis added).

James's suggestions are however by no means fully conclusive. Whitehead observes that James has not fully disambiguated the distinction of entity and function, since in one way 'the notion of "entity" is so general that it may be taken to mean anything that can be thought about' (SMW: 144). If the primary obligation of empiricism is taken to be establishing definitive epistemic criteria, then pure experience seems fated to entail a subjective relativism in any practical sense. Even if there are no longer two metaphysical substances to be traversed, because the 'same' element can be determined in different ways from different perspectives, the practical consequence is a subjective relativism. This has been a major concern and it remains the case that problematising fixed identity is often conflated with relativisms or anti-realisms.[35] To see the radical empiricist shifting of the terms of the problem (from substances to processes and events) as a realism, we have to examine pure experience as a metaphysical posit, not exclusively a phenomenological device. This posit intensifies the intersection between lived experiences of individuality and abstractions through which this experience is conceptualised.

Realism and asubjective experience

James's description of pure experience as 'primal stuff . . . of which every-thing is composed' is apt to be misunderstood in two ways. The first fol-lows the extent to which the language of 'stuff' is easily read through a substance metaphysical paradigm. Consider Bertrand Russell's description of pure experience as a 'neutral monism': 'James . . . advocate[s] . . . "neu-tral monism", according to which the *material* of the world is constructed neither of mind or matter, but something anterior to both' (Russell 1961: 767–8, emphasis added). On the one hand, Russell's description of pure experience as prior to mind and matter *logically* is clearly correct and con-sistent with implications also noted by David Lapoujade's Deleuze-inspired reading. In beginning from 'a field in which experience is virtually subjec-tive or objective, indifferently mental or physical, but also primitively nei-ther one nor the other', pure experience is free from 'the categories with which it is traditionally partitioned' (Lapoujade 2000: 193).[36] But every-thing depends on how we understand the status of this monism. Russell's language of 'material' is easily assimilated into an underlying metaphysics of substance. While he is right that pure experience is anterior to both mind and matter, he still suggests this primal stuff as a kind of 'third substance' – some unforeseen prime element. This does not track the way James charac-terises 'stuff' in terms of relational events. In this regard, it would be more accurate to describe pure experience as do-ing or happen-ing prior to the sorting of happen-ings into contrasting types: mind or matter.

The second misunderstanding hangs on the language of neutrality and purity. Does neutrality or purity name a universal and homogeneous essence? Purity understood in this sense encourages a collective approach where pure experience is the set of all the pure elements that make up a reality finished and closed. The neutrality and purity of this monism entail a homogeneity at the 'elemental' level. This collective approach is contrasted with a distributive approach that understands pure experience as open and to a certain extent contingent. The 'elements' of pure experi-ence do not share a common qualitative form and are not determined in a single 'neutral' way. Their purity does not name a universal homogeneity, despite the language of 'primal stuff'.

James has this latter sense in mind when he stresses '. . . there is no *gen-eral* stuff of which experience at large is made. There are as many stuffs as there are "natures" in the things experienced' (1976: 14, emphasis added). James denies collective homogeneity: pure experience is not an undif-ferentiated goop. This distributive notion emphasises open contingency. Events of pure experience do not necessarily follow a single principle of determination that can be ascertained outside of pure experience itself. This does not resolve epistemic or existential problems of differentiation, but it does establish a form for the problems. Accounting for individuated differences is not a single problem of designating connotation or picking out a particular element. Instead of one way to designate, there will be many, and they will depend on the relations and 'natures' in question.[37]

Pure experience thus intensifies the extent to which activity of identification is bound up with underlying metaphysical assumption. James's descriptions of pure experience in relation to 'newborn babes' and 'men in semicoma' (1976: 45) is misleading if it encourages a psychological conception only – as if pure experience were an indiscriminate chaos that results from failure to organise reality. Such a psychological read begs the question in presuming the ordering of pure experience happens automatically and universally. This presumption is enabled when pure experience is merely an indeterminate flux full-stop. The metaphysical question of differentiation thus hangs on the status of potential and actual relations inherent in pure experience, with identification and differentiation akin to processes of actualisation.

It is true that pure experience presents an immediate flux in which there is not yet subject or object *as such*. However, to say that it is 'pure in the literal sense of a *that* which is not yet any definite *what*' (James 1976: 46) does not mean it lacks any direction or relation whatsoever. Instead, pure experience is seething with difference and relation, but these relations have not been ordered categorically. *These relations are not yet determinate, but they are determinable.* They are 'ready to be all sorts of whats; full both of oneness and of manyness' but they are not yet 'caught' (James 1976: 46). The 'prepositions, copulas, and conjunctions' that 'flower out of the stream of pure experience' and then 'melt into it again' are not arbitrary inventions. It is not that anything goes, because pure experience is not a neutral homogeneous whole but rather an in-process network of relational moments. Lapoujade alludes to this using Deleuze's 'patchwork' metaphor, 'The textile matter of pure experience is composite . . . it consists of fragments linked to each other in different ways' (2000: 196). A metaphysics of pure experience thus affirms an empiricist orientation: because experiential events are connected through active relational transitions that remain contingent, there is no substitute or replacement for continued attention to the events of experience.

This is not *an answer* to the question of determination. As both Whitehead and Deleuze develop it, the metaphysics of differentiation requires going beyond the constraints of a narrow empiricism. The question though is how. Rather than reductive opposition between rationalist and empiricist, with both representing opposing 'positions' complete in themselves (either universal predetermined principles of identification or identities given *in toto* in experience), there is need of *both* principles and experience. But this coupling of principles and experience is itself relationally emergent and contingent. This is why emphasising pure experience as a metaphysical posit exceeds the explicitly phenomenological. The determination of the object or identity in question is not simply given at the level of pure experience because determination is a processual event. The crucial innovation is *pure experience's determinable but not determinate distinction*. This distinction is not only epistemic – if it were, we would be back at the psychological reading in assuming that objects in themselves are determined ontologically, but not at the psychic level. Pure experience would then refer only to pre-processed perceptual information before achieving a cognitive representation matching the objective determination in reality. I am suggesting that this *determinable but not determinate distinction is metaphysical, not only psychological. In this sense, determination as actualisation refers to ontological processes that exceed human perception alone.*

Emphasising pure experience as metaphysical posit with a determinable but not determinate structure sets basic criteria for the projects of both Deleuze and Whitehead. As already stressed, because events of pure experience cannot be determinatively prefigured, these projects remain empiricist in orientation. This orientation is particularly relevant for thinking existential effects. While both Deleuze and Whitehead exceed empiricism narrowly construed in constructing metaphysical concepts that cannot be directly verified, these concepts never preclude the need for continued attention to a reality that remains open and contingent. This existential orientation is correlative with what Deleuze calls 'the secret of empiricism'. Rather than 'simple appeal to lived experience', empiricism is a 'mak[ing], remak[ing] and unmak[ing]' of concepts 'along a moving horizon': 'the concept as object of an encounter' not a universal *a priori* (DR: xx).

Reading the determinable but not determinate structure as metaphysical – not only psychological or phenomenological – introduces the necessity of a transcendental level, in roughly but not precisely the Kantian sense of conditions for the possibility of experience. Most notably, for neither thinker is it necessary for this transcendental level to be exclusively threaded through the locus of a determinate human subject or ego, whether empirical or transcendental. While Whitehead acknowledges Kant's importance in 'introducing . . . the conception of an act of experience as a constructive functioning', he inverts the direction of Kant's analysis (PR: 156). Whitehead writes: 'for Kant the process whereby there is experience is a process from subjectivity to apparent objectivity' (PR: 156). Kant's transcendental deduction thus seeks to discover the necessary conditions that govern the construction of objective experience.

In doing so, Kant 'presupposes a subject which then encounters a datum . . . and then reacts to that datum' (PR: 155).[38] For Whitehead, however, there is not a pre-existent subject constructing an objective experience, there is, rather, a real external and objective world *out of which* a process achieves subjective experience. He thus seeks to 'explain[s] the process as proceeding from objectivity to subjectivity, namely, from the objectivity, whereby the external world is a datum, to the subjectivity, whereby there is one individual experience' (PR: 156). The function of the transcendental is thereby shifted. Though Whitehead still seeks to think how reality must be to explain subjective experience, his categories, both transcendental ('eternal objects') and actual ('actual occasions'), are not restricted to human perspective alone.

Whitehead's inversion bears some similarity to what Deleuze refers to as a 'transcendental empiricism' that seeks to uncover the 'conditions of real experience, and not only of possible experience' (DR: 285). Such transcendental empiricism also departs from the presupposition that these conditions must go through a unified transcendental subject. Both Deleuze and Whitehead see Kant as reducing experience to representational cognition, precisely because, as Whitehead puts it 'Kant's act of experience is essentially knowledge [such that] whatever is not knowledge is necessarily inchoate and merely on its way to knowledge' (PR: 155). This emphasis on knowledge leads Kant to only grant extensive and representative features of experience and privilege the form of identity as a necessity for coherence. This denies pervasive features of real experience: sensible, affective and intensive qualities that Deleuze refers to as 'the lived reality of a sub-representative domain' (DR: 69).[39]

By only granting extensive and representative features of experience as those that lead to 'knowledge', Kant's tracing of the conditions of possible experience is not *truly transcendental*. This is why Deleuze states that he seeks conditions of real rather than possible experience. If the conditions of real experience are traced off cognitive representations, as in Kant, then there is a risk of a reductive circle that blocks the full intensity of sensible encounters by ensuring that their results must fit into categories of extensive representation. This allows a perceived epistemic requirement (and Deleuze is most worried about the necessity of identity over time) to constrain what we grant in experience. It hinders attention to the real by prejudging what counts as viable and results in a failure to respond to what is primary in experience, namely, difference:

> Empiricism truly becomes transcendental . . . only when we apprehend directly in the sensible that which can only be sensed, the very being *of* the sensible: difference, potential difference, and difference in intensity as the reason behind qualitative diversity. (DR: 56–7)

Understanding the determinable but not determinate structure as metaphysical shows the constructions of Deleuze and Whitehead to be varieties of nonstandard realisms, provided we accept a gap between ordinary representations of reality and processes constitutive of these representations. Such a realism does not map onto affirmations of the objective as statically given. Both construct metaphysics that are realist in rejecting the necessity of a human subject, but also perspectival. This perspectivism is not only epistemic, it is rather a feature of the genesis of the real – what Deleuze would characterise as a 'differential' genesis. In this way, we might think of their work as affirming objective relativism provided we relinquish the notion of the human subject and the material object as stable and fixed pivots. This also requires giving up the idea that experience is necessarily a predicate of an experiencer.

Such moves challenge habits of lived individuality. What could it mean to affirm lived experience not necessarily tied to an experiencer? This question requires working to understand the conceptual innovations necessary for its coherence *and* translating these innovations into terms of living experience –while remaining mindful that available language for this translation is in pervasive tension with the vision being articulated. We are now better prepared however to understand why someone would risk this challenge, as well as some of its first steps. Avoiding a bifurcation of nature is not primarily a matter of resolving the epistemic status of different kinds of knowledge, but rather a project of creating conceptual conditions by which we might learn to rehabilitate living experience. If such creation remains speculative and experimental under current conditions, its goal is not idle speculation. As Stengers puts it, metaphysics is 'an experimental practice like physics . . . it devises concepts that will have no meaning unless they succeed in bringing into existence those dimensions of experience that usual statements can ignore' (Stengers 2011: 248). The speculative concept must connect with reality even as it challenges ordinary habits of perceiving that reality.

Notes

1. I cite Deleuze's reference to these dialectical turns without assessing their legitimacy.
2. Kant repeats this phrase no less than four times in the first two pages of the B Preface alone.
3. Whitehead's 'propositional lures' or 'lures for feeling' reflect how different ideals inform the ongoing constitution of actual reality. Such 'lures', which may be conceptual, ethical, emotional, and so on, help to order the process by which an actual occasion becomes determinate (PR: 185–6).
4. Whitehead's full quotation reads: 'we habitually observe by the method of difference. Sometimes we see an elephant, and sometimes we do not. The result is that an elephant, when present, is noticed' (PR: 4).

5. Emphasis on the problem extends from Deleuze's first monograph on Hume through the metaphysical texts of the late 1960s and the co-authored work with Guattari. His reading of the history of philosophy is oriented around problems rather than positions, in contrast to 'histories . . . in which solutions are reviewed without ever determining what the problem is . . . since the problem is only copied from the propositions that serve as its answer' (WIP: 80).

6. This paragraph closely follows Le Doeuff's reading in *The Philosophical Imaginary*.

7. Given Kant's socio-political context of increasing European colonialism as well as his prominence in articulating racist hierarchies as justification for such conquest, we might well ask if Kant's language of possession or his scorn for those useless nomads are easily dismissed as merely inept metaphors.

8. Kant's privileging of work over play indicates a valuation that is not obviously universal. Deleuze's own inclusion of puns and wordplay are a performative challenge to this image of 'real' thought, which, as he reminds us, is a moral image. This tension remains in the norms of professionalised philosophy and Deleuze alludes to it discussing Whitehead's relative lack of attention in 1987: 'Whitehead is read by a handful of enthusiasts and another handful of professionals. Like Bergson as well . . . one is unable to say whether or not it is "serious" stuff (http://www.webdeleuze.com/php/texte.php?cle=140&groupe=Leibniz&langue=1 (my translation)).

9. Whitehead's schools resonate with P. F. Strawson's well-known distinction between 'revisionary' and 'descriptive' metaphysics. Where the former 'is concerned to produce a better structure [of thought about the world]', the latter wants only to describe the currently existing structure. Strawson observes that this distinction is imperfect insofar as 'no actual metaphysician has ever been, both in intention and effect, wholly the one or the other' (1959: xiii). That said, both Whitehead and Deleuze clearly emphasise the revisionary.

10. Compare this with Whitehead's claim in PR: 'in the real world it is more important that a proposition be interesting than that it be true' (PR: 259). This is further developed in *Modes of Thought*, where Whitehead focuses on Importance, Interest and Perspective to complicate positivistic epistemologies. See especially MT: 1–19, 21.

11. Whitehead distinguishes between Aristotle's own philosophy and the influence of his logic: 'The dominance of Aristotelian logic from the late classical period onwards has imposed on metaphysical thought the categories naturally derivative from its phraseology. This dominance of his logic does not seem to have been characteristic of Aristotle's own metaphysical speculations' (PR: 30).

12. This terminology comes from Guattari's *The Three Ecologies*. See Chapter 5.

13. This worry (along with an insistence on a framing opposition between idealism and materialism) drives a constellation of critiques of Deleuze, especially Hallward 2006 and Žižek 2004. Badiou's critique of Deleuze as a 'thinker of the One' is also in the same conceptual universe.

14. This inaugurates Whitehead's habit of naming constitutive fallacies: the 'fallacy of simple location' (SMW) 'fallacy of misplaced concreteness' (SMW) the 'evolutionist fallacy' (FR) and so on.

15. Stengers follows Lewis Ford in locating a leap in Whitehead's thought between *The Concept of Nature* (1920) and *Science and the Modern World* (1925). Deleuze's most extended remarks on Whitehead, both in print (*Le Pli*) or in his 1987 seminar, draw extensively on *The Concept of Nature*. It is worth nothing that this text did not appear in French until 2006. During Deleuze's life, the following works were available in French translation: *Science and the Modern World* (French version published in 1930); *Religion in the Making* (French version published in 1926); and *The Function of Reason and Other Essays* (French version published in 1969). This last text includes what is published separately as *Symbolism* in English, as well as *Nature and Life*, two lectures also included in *Modes of Thought*. It appears likely that much of Deleuze's access to Whitehead would have come through Jean Wahl, who unlike Deleuze was fluent in English.

16. This is not to say that Whitehead's discussion of events in *Concept of Nature* is equivalent to his metaphysical construction of the actual occasion. Debaise argues that conflation of 'actual occasion' as equivalent to an experiential event stems from Deleuze (Debaise 2017b: 54–5, 62 n13; for an example, see Shaviro 2014). To be sure, in his most extended engagement with Whitehead in *The Fold*, Deleuze does not acknowledge this distinction, solely using the term 'event' (1993 [1988]: 76–82).

17. Whitehead's criticism of conceiving matter as an entity discretely located in space and time runs throughout his work. His criticism makes his work relevant to 'new materialisms' of recent years, many of which take inspiration from Deleuze. As Diana Coole and Samantha Frost write, one aspect of this new materialism is its rejection of matter imagined as a 'massive opaque plenitude' and instead a conception of matter as 'indeterminate, constantly forming and reforming in new ways' (2010: 10). However, such new materialisms, many of them feminist in orientation, wrestle in much more substantive fashion with the political implications of this conception of matter, especially for lived phenomenological categories such as gender.

18. Whitehead's bifurcation can be seen as a precursor to what has come to be known (following Chalmers) as the 'hard problem of consciousness' in contemporary philosophy of mind. For readings that engage this question using Whitehead see Shaviro 2014: 65–84; Basile 2018: 47–60.

19. Deleuze makes a similar point about the inescapeability of abstraction: 'True lived experience is an absolutely abstract thing . . . I don't live representation in my heart, I live a temporal line which is completely abstract. What is more abstract than a rhythm?' (Cited in Atkins 2016: 355).

20. We might say that the 'event' is the nontechnical formulation of a basic shift and the 'actual occasion', as we will see, is the technical construction it motivates.

21. Examples include Debaise's *Speculative Empiricism* 2017a; also Alliez 2004, Robinson 2009. Debaise notes the relation of 'speculative empiricism' with radical empiricism and pragmatism (2017a: 7–10) a point shared by Massumi's usage of a 'speculative pragmatism' (2002, 2014). Besides James, varieties of radical empiricisms can also be found in Bergson, Peirce and F. W. J. Schelling, among others. The emphasis on empiricism departs from a rationalist emphasis in the tradition of process theology heavily influenced by Hartshorne. It is also by no means a consensus view in Deleuze scholarship. See Bryant 2008 and Kerslake 2009 for studies emphasising Deleuze's complicated relationship with Kant and post-Kantian idealism rather than empiricism.

22. I refer to Beckett's quip, since become cultural vernacular: 'Try again. Fail again. Fail better' (1989: 89).

23. This is not intended as exclusive antecedent. Though James is important for both, his influence is by no means fully decisive.

24. One exception to this tendency to focus on the epistemological and methodological rather than the metaphysical is found in David Lamberth (1999).

25. This tradition was particularly prevalent in the emergence of psychology as a discipline, in which James was a leading figure. For a discussion of the Associationist tradition, see Young 1968. For an appraisal of James's psychology as a response, see Klein 2009.

26. Whitehead adopts this Jamesian phenomenological critique in the *Symbolism* lectures but complicates the continuity of experience by distinguishing two modes of perception: causal efficacy and presentational immediacy (S: 31–2). This text is notable as a transition between the phenomenological and metaphysical. See also Faber et al. 2017.

27. Vicious intellectualism refers to a reification of conceptual distinctions that is then used to restrict experiential possibility: 'The misuse of concepts begins with the habit of employing them privatively as well as positively, using them not merely to assign properties to things, but to deny the very properties with which the things sensibly present themselves' (James 1977: 99).

28. Stengers describes Whitehead as desiring to confer upon James's suggestions a more 'rigid coherence' in thinking through their implications 'right to the end' (Stengers

2014: 150). She cites Whitehead's remark that 'Every philosophical school . . . requires two presiding philosophers. One of them, under the influence of the main doctrines of the school, should survey experience with some adequacy, but inconsistently. The other philosopher should reduce the doctrines of the school to a rigid consistency' (PR: 57).

29. In Deleuze's words, his teacher Wahl: '. . . not only introduced us to an encounter with English and American thought, but he had the ability to make us think, in French, things that were very new' (1987: 57–8). A significant figure in French philosophy in the early to mid-twentieth century, Wahl's eclectic vision melds an existential orientation informed by Kierkegaard and Heidegger, German Idealism and Romanticism, and radical empiricism and process metaphysics. Recent years have seen renewal of interest in his work and two new translations. See Wahl 2016, 2017. For discussion of the influence on Deleuze, see Bowden et al. 2014; Madelrieux 2014; Zamberlin 2006.

30. This is presented most famously in the *Phenomenology* (see Hegel 1977 [1807]: 58–79).

31. This text was published in English translation in 1925 as *Pluralist Philosophies of England and America*.

32. In addition to departing from the emphasis on pragmatism as epistemological method, Wahl insists that radical empiricism leads to a realist pluralism. This departs from tendencies to read James's pragmatism as either leading to, or following from, what is now called 'anti-realism'. See, for example, Seigfried 1990.

33. This text, with a new preface by French scholar of pragmatism Mathias Girel, was reissued in 2004 by J. Vrin.

34. This interest follows Bergson's critique in *Matter and Memory* of the way in which realisms and idealisms mirror each other in their presuppositions.

35. One trend within pragmatist scholarship is to portray this as a split between the 'realist' pragmatism of Peirce and the bad 'subjectivism/relativism' of James. See for example Misak 2013 (especially 51–73) who narrates pragmatism as a split between the hero Peirce and the well-intentioned but confused James.

36. A reading seconded by Jeffrey Bell, who sees James's pure experience as analogous to Deleuze's virtual (2009: 21).

37. This distributive approach also displays James's manner of distinguishing between empiricism and rationalism: 'empiricism means the habits of explaining wholes by parts, and rationalism means the habit of explaining parts by wholes' (James 1977: 9). If identity is thought of as a conceptual whole, then the implication is that its qualities are not given in the concept but rather we must begin with these qualities (relations) in seeking to characterise the concept. This is why, for example, Deleuze states that 'Relations are external to their terms. When James calls himself a pluralist, he does not say, in principle, anything else' (Deleuze 1991: 99).

38. The datum corresponds to Kant's *noumenal* thing-in-itself (*Ding an sich*).

39. Observing that 'Deleuze's most general problem is not being but experience', Zourabichvili explains that 'experience' does not designate only ordinary lived experience: 'Transcendental empiricism means first of all that the discovery of the conditions of experience itself presupposes an experience in the strict sense: not the ordinary or empirical experience of a faculty (for the data of empirical lived experience doesn't inform thought about what it can do), but this faculty taken to its limit, confronted by that which solicits it in its own unique power' (Zourabichvili 2012: 210).

Chapter 2

Individuation and Attunement: Identities in Process

Pure experience intensifies the question of individuation.[1] Lacking a transcendental subject as source of identity, its perspectivism raises questions about how to stabilise and identify objects or enduring subjects. Both emerge within events of relation even as these events intersect in different series (and hence suggest no single unitary perspective as the sole real). This chapter explores how a process metaphysical approach shifts the approach to these questions. After briefly reviewing individuality as a metaphysical problem in general, I show how a process approach changes its framing.

Constructing a problem: individuality at the intersection of the abstract and existential

Despite the apparent ubiquity of individuals in experience, cogent understanding of individuality is problematic. On the one hand, there is a strong appeal to its intuitive self-evidence: what is more obvious than individual difference – the difference between you and I, the difference between this day and that, the particularity of *this* thought, *this* feeling, *this* moment? But an ability to identify and especially communicate such individual moments or experiences seems to require appeal to generality that exceeds the individual. If something were truly individual, would it not be beyond communication or representation? Must an individual be singular? If not, what does it mean to call something an individual? What is the status of its individuality in relation to properties that seem, by virtue of their knowability, to be general and shared?

The puzzles implied in these questions can seem contrived. Worse, when a conceptual system resolves them through denying the existence of individuals, philosophy can appear unduly scholastic in a pejorative sense. Whitehead expresses this worry in relation to Bradley's Absolute:

> The point to be emphasized is the insistent particularity of things experienced and of the act of experiencing. Bradley's doctrine – Wolf-eating-Lamb as a universal qualifying the absolute – is a travesty of the evidence. *That* wolf ate *that* lamb at *that* spot at *that* time: the wolf knew it; the lamb knew it; and the carrion birds knew it. (PR: 43)[2]

And yet, while undoubtedly the experience of that wolf and that lamb at that spot and time was a surprise for each, Whitehead's affirmation of '*the evidence*' rests on imaginative projection that challenges compartmentalised notions of individuality. Presumably neither author nor reader have had the precise experience of that wolf or that lamb. If an existentially viable metaphysics cannot ignore the lived particularity of individualised difference, this still does not mean it can presume self-evidence for *thinking* individuality. Individuality *is* a philosophical problem. The question is orienting an engagement that does not lose touch with experience.

To begin, this means not dismissing the juxtaposition of lived ubiquity and conceptual puzzlement as irrelevant, nor presuming a successful 'solution' means eradicating this strangeness through definitive conceptual closure. Rather, this *strangeness* of the ordinary indicates a kind of 'encounter' that Deleuze identifies as necessary for 'forcing us to think' ('Something in the world forces us to think'). The *feeling* of strangeness generated by scrutinising an apparently obvious feature of experience expresses something important about the existential condition of an open reality. Rather than denying this strangeness, it is a clue towards understanding individuals as contingently emergent expressions of processes rather than closed final entities. Within such a processual reality, individuals are by degrees and individuation is an ontological process rather than an exclusively epistemic or descriptive one.

Developing these claims requires understanding how they depart from assumptions that typically frame study of metaphysical individuals.[3] Jorge G. E. Gracia's exhaustive survey of the metaphysics of individuality is exemplary for the skill with which it pursues a result situated within this framing. However, despite introducing significant nuance in distinguishing six different issues in 'individuality' (intension, extension, ontological status, principle of individuation, discernibility, and linguistic reference), his study still maintains constitutive features that a process approach reconfigures or rejects.[4] These are: (1) universals and particulars as the exhaustive categories for considering individuals; (2) a strict logic of exclusive disjunction and rigid law of excluded middle; (3) a presumption that macro-level representations of apparently self-identical objects are paradigmatic for how we conceptualise 'individuals'; and (4) a background premise of a-temporality. The first two features are formal and the second two are better described as lived or experiential.

Formally, the distinction between universals and particulars is not easily evaded, since it is within this framework that the metaphysical issue of individuals historically and conceptually arises. Consider my opening example, where the problem involves understanding the way that qualities (presumed 'universals') as predicates relate to their bearers (as presumed particulars). Gracia's study accordingly designates the intension (conceptual definition) of the individual as a 'noninstantiable instance', that is, a particular that cannot itself be re-instantiated in another particular.[5] The metaphysical category of individual appears dependent on this structure, particularly within the canonical history of European philosophy as the terrain of the medieval debate between so-called 'Realists' and 'Nominalists'.

This relation between these two categories is most often treated through a logic of exclusive disjunction. Gracia writes in reference to the universal and particular that 'Everything in the universe is either one or the other, and if it is one, it cannot be the other' (1988: 59–60).[6] As we will see, the process-inflected works of both Whitehead and Deleuze are unthinkable through a rigidly held law of excluded middle. Both utilise logics of degrees, intensities and included middles. Insufficient attention to the nuances of this point is a source of caricatures – whether overenthusiastic or prematurely dismissive. While both are critical of inflated conceptions of the power of formal analysis for constructing an adequate metaphysics, this criticism is never simply rejection or denunciation. To say, as Whitehead does, that philosophy has functioned with 'a false estimate of logical procedure with respect to certainty' is not to dismiss all rational criteria or rigour.[7] The issue has to do with the extent to which operative formal principles can become reductively hegemonic. This reduction is frequently a matter of misapplication where the scope of a sound formal principle is uncarefully expanded in a manner that reduces or dismisses aspects of experience that cannot be coherently encountered from within the confines of the principle. This demonstrates a mistaken order of operations. As Whitehead puts it, 'logic presupposes metaphysics' not vice versa (MT: 107).[8]

At issue is what form a logic takes and how it is applied. Nicholas Rescher has noted that a consistent process metaphysics requires a nonclassical logic. If we take the central kernel of a process approach to be the idea that any individual 'x' is constituted metaphysically as the outcome, *in varying degrees of stability*, of actions or processes, then we have to relinquish or at least loosen the laws of noncontradiction and excluded middle.[9] A shift to process:

require[s] the deployment of two historically unorthodox items of concept-machinery: (i). A 'fuzzy logic' – or at any rate a fuzzy mathematics – that puts the conception of indefinite (imprecisely bounded) intervals and regions at our disposal; (ii). A semantics of truth-value gaps, serving to countenance propositions that are neither (definitely) true nor (definitely) false but indeterminate in lacking a classical truth value. (Rescher 1996: 67)[10]

The emergence of non-classical formal logics over the last fifty years shows that neither Deleuze's nor Whitehead's worries about classical logic must necessarily result in unbridled irrationalism or anti-intellectualism.

For both, a process approach to individuation depends on logics of degree and intensity that complicate reductive binarisations. As we have seen, the formal operation of exclusive disjunction is frequently conflated with metaphysical dualism. Rather than discrete 'all-or-nothing' binary switches, logics of intensity follow tendencies or orientations within processes. Processes may partake of divergent polarities in different degrees, but such polarities cannot be fully extricated or separated out as pre-existent and discrete metaphysical kinds. This logic of included middle is necessary to understand Deleuze's habit of presenting series of oppositions: the smooth and the striated, the molar and the molecular, the continuous and discrete, the problematic and the axiomatic, the intensive and the extensive. Brent Atkins shows that rather than categorical differences in kind, such pairs express differing tendencies: for example, the intensive as 'tendency toward change' and the extensive as 'tendency toward stasis' (2016: 354). Both are asymptotic limit conditions: complete flux or maximal stasis are never achieved, and we do not switch from one to the other all at once. A logic of intensity is also inherent in Whitehead's claim that 'each actuality is essentially bipolar, physical and mental, and the physical inheritance is essentially accompanied by a conceptual reaction' (PR: 108). Whitehead incorporates into the actual occasion the two poles (physical, mental) so that all occasions already include both, but in varying degrees of intensity.[11]

Such formal considerations challenge lived assumptions pervasive in standard framings. Taking a presumed stable and self-identical object (what Deleuze calls 'something=x' or 'object=x') as the experiential paradigm of individuals risks begging the question, *especially when coupled with background premises of a-temporality and a-relationality*. For one, the stability of such an entity is clearly an abstraction. Identification and individuation of even a bare object partakes of significant contingencies: cultural custom, linguistic convention, physical temperature and pressure modulated within sustainable intensities. While it is possible to dismiss these as 'empirical only', the issue is the extent to which standard analyses work backwards from experience of such presumed individual objects to establish essential features of a metaphysical category. The category is thus conceived, by definition, as a-temporal and a-relational.

In addition to failing on radical empiricist grounds, this displays what Deleuze terms a 'transcendental fallacy' whereby the form of the individual at the empirical level is presumed as the necessary form at the transcendental level. Rather than offering a genetic account of how individuation emerges, they 'give up genesis and constitution and . . . limit [themselves] to a simple transcendental conditioning' (LS: 105). This assumes that the empirical and the transcendental have a formal relation of *resemblance* and *representation*: 'The error of all efforts to determine the transcendental as consciousness is that they think of the transcendental in the image of, and in the resemblance to, that which it is supposed to ground' (LS: 105).

Whitehead and Deleuze instead consider *processes of individuation* without presuming that the form of these processes is traced off the apparent existence of objects as entities. This is a different construction of the problem, not a final solution. Given the experience of individuated difference, for a process account of individuation to remain existentially relevant it cannot collapse or dismiss all individuation into process or flux. And yet, their account is under a different obligation: what individuation looks like changes if the 'item' to be individuated is conceived as verb rather than noun.

The question is no longer what makes an individual (whether object, subject or abstract category) the individual x it is, but rather why some events repeat in relatively sustainable or recognisable patterns and others do not. The question of individuals becomes more one of consistency than strict identity.[12] If from within the standard problem's paradigm this denies the existence of individuals, this denial cannot be assimilated without remainder into scholastic or neo-Platonic realisms or absolute idealisms. Rather, both Deleuze and Whitehead split the difference between radical nominalisms and quasi-Platonic realisms by affirming the singularity of individual moments or *haecceities*, while also acknowledging that a coherent account of individuation requires a transcendental structure to explain how processes or events consolidate into macro-stable objects or subjects. Call these the empiricist and rationalist (or 'transcendental') sides of their projects. Both sides are necessary.

Additionally, both agree that individuation must consider temporality because what it is to be an individual, if there is such a thing, is to endure in a stable pattern over time. Individuating experience is experiencing that coheres or repeats through a duration. Insisting on its temporality deepens the entanglement of the abstract and the lived. Though the abstract problem of individuals applies to all self-identical entities, personal or subjective identity is where this scholastic question takes on greatest existential weight. Indeed, since Plato but especially in the early moderns, the question of personal identity is inextricable from temporal questions of recollection and memory, as for example in Locke.[13] Moreover, introducing temporality into processes of individuation results in a pluralising of the existential individual. Though technically oxymoronic (by definition an individual is that which is not plural), existentially this responds to a poetics of multiple selves expressed in the Whitmanian 'for I contain multitudes' or the Rimbaudian 'Je est un autre'. The question again is how and why certain multitudes cohere and what it means for possible coherence or incoherence to be *lived*.

The lived subject is thus bridge between metaphysical individuation and lived experience.[14] Many images of metaphysics bracket this bridge by focusing on a presumed discrete object as paradigmatic of an individual existent. And yet, the conceptual question of individuation is *lived through* subjectivity, both in terms of the subject's experience of individuated difference and in their *conception of* subjectivity. This is not to say that the metaphysical question of individuation is equivalent to what Deleuze and Guattari describe as 'subjectification'.[15] Individuation is a broader issue than subjectivity as such. But processes of subjectification operate within this broader technical context of individuation. I therefore concentrate here at a more technical level to emphasise *the difference* between metaphysics and psychology. Insisting on this difference will be necessary when we turn to Guattari's account of 'psychic ecologies'. It is also important because of the extent to which a processual account of individuation challenges habits of description frequently conflated with directly perceived realities. Everything depends on understanding the process alternative as not simply descriptive and yet still responsive to the real. Its conceptual innovations cannot be justified through phenomenological appeal alone because they are situated at the intersection between living reality and ordinary *conceptions* of it. Neither Deleuze nor Whitehead are doing phenomenology, and yet taking their metaphysics seriously induces phenomenological effects especially regarding the stakes of habits of attention. It is these effects I am interested in. Accordingly, my goal in what follows is less to offer a conclusive account of individuation and more to set up the relevant questions, implications and consequences for the intersection of affect, attention and habits of subjectivity.

Three theses on individuation in process thinking

Thesis 1: individuality and identity are not the same thing

This distinction is not unique to a process approach, but such an approach shows why it matters. In Gracia's schematic, the distinction is between extension and discernibility: individuality is a metaphysical status and has to do with extension (are there any things that are individuals?) and identity is an epistemic criterion of discernibility (that x there is an individual).[16]

Casual use of language blurs this distinction. In *The Lies that Bind: Rethinking Identity*, Kwame Anthony Appiah discusses how pervasive use of what linguists call 'generics' reinforce psychological habits of 'essentialism'. A 'generic' refers to an unspecified or unbounded proposition. Appiah gives the following as examples: 'Tigers eat people' or 'Mosquitoes carry the West Nile virus' and observes 'it's . . . hard to say what makes generics true' (Appiah 2018: 26–7). For example, in the case of the mosquito claim, 'an epidemiologist can sincerely [make this claim] while knowing that 99 percent of [mosquitoes] do not carry it' (Appiah 2018: 27). Conceptually, it is easy to see that 'generics' require more precision with the addition of a qualifier: Some . . . A few . . . All . . . No . . . and so on. Though such qualification makes the claim more precise, it does not necessarily alleviate what Appiah identifies as the biggest challenge: we are more likely to accept a generic if 'the property it mentions is one that we have a reason to be concerned about' (Appiah 2018: 27). Knowing that mosquitoes carry West Nile virus makes us scared of mosquitoes, even if most of them do not. Though the majority of tigers probably have not, and will not, eat a person, if we encounter a tiger, this generic claim shapes our behaviour in consequential ways.[17]

In addition to the extent to which generics do or do not trigger concern (whether fear based or otherwise), they are also reinforced by a metaphysical habit. That is, we are more likely to accept a generic uncritically if 'we think of the class (tigers . . . mosquitoes) as a kind, a group . . . with a shared essence' (Appiah 2018: 27). Psychological research suggests that this habit of assuming a social classification or category names 'an inner something . . . that explains why [members of the classification] have so much in common' appears as early as two years old.[18] The temptation, given this early appearance, is to think of this as an innate attitude, and from there to think of it as following something given in reality. But we should not assume that psychological habits are insulated from metaphysical habits, especially those built into the grammar of language. Indeed, words as such are predisposed to perform this generic habit, functioning as modes of grouping different particulars into a broader category. But they can be used in more or less skilful ways. Essentialism as a psychological habit depends on an implicit modelling of groups as referring to underlying static substances. This is a metaphysical habit – not a given. Appiah shows that this habit is *not merely abstract* and not easily evaded.

Though there is a metaphysical dimension to this habit, it is so entrenched that mere description of an alternative metaphysics is insufficient for altering lived habits. Rather, more thought must be given to how metaphysics infiltrates habits of attention. The entanglement is complex. Unreflective deployment of 'generic' statements is reinforced by an essentialist understanding of groups that is to some extent embedded in language as using general markers to refer to particulars. These inform habitual inferences undergirded by a substantialist metaphysics. This entanglement orients attention in the service of inductive predictions, demonstrating a loop between experience and concepts. The speculative gambit of this text involves experimenting with how alternative concepts might encourage different loops. This is not an all-or-nothing switch, either experientially or conceptually. Nevertheless, because identities do not name static essences, a process approach to individuation more starkly draws attention to the difference between a generic identity ascription and processes generative of individuals. Exploring how serves as propadeutic for a different prioritising of attention. Conflation of identity and individuality frequently generate expectations reinforced under prevailing substantialist assumptions. How does a processual approach to individuation invigorate a different construction of what is relevant and important?

For both Deleuze and Whitehead, macro-level identities name manners of action or process, not static entities.[19] Such manners may be grouped into relevant categories, provided we bracket the assumption that their source is an underlying entity. This problematises rigid identity designation, since results of these processes may alter over time. This can shift habits in describing perceptual encounters because their outcome need not be structured entirely by identity ascription. If identity is not fully definitive of individuality, identities become retroactive ascriptions of events or processes. We move from: that is a car, those are Americans, that is a white-crowned sparrow (*Zonotrichia leucophrys*) to red motion past window now, American English speaking amid Czech on trolley car, sharp whistling sound into trilling notes, flash of grey brown under the bush. This encourages more intense attention to the specificity of the event in question and unsettles assumptions that macro-level identity ascription is explanatorily sufficient for the individual action in question.

The white-crowned sparrow sighted in the example above is not created by its observation by a human perceiver in some kind of subjective idealism.

Such worries are mediated through the assumption of the 'individual' as necessarily designating an object. Something denoted by the term sparrow exists even when not being observed by the human. But the alternative I am bringing into view changes how we understand what 'the sparrow' refers to. Rather than a static object designated in a one-to-one correspondence with a static designator (it=sparrow), a white-crowned sparrow is more properly speaking a 'sparrowing' that inflects occasions in a certain manner that repeats in different contexts.[20] This inverts ordinary habits of predication. Sparrow-ing is not a description of a being, it is the manner that constitutes what we come to identify (through procedures of abstraction and inductive generalisation) as a species.

This is a radically different way of thinking about identity *and* individuality. Given the extent of its departure from ordinary modes of orientation, I will first offer additional motivations through a more formal discussion, in this case Whitehead's unorthodox discussion of how 'variables' function in logical reasoning. Ordinarily the syntactic variable x is presumed to maintain a self-identity independent of context. But, as Whitehead points out, 'complete self-identity can never be preserved in any advance to novelty' (MT: 107). The syntax of symbolic reasoning covers over reckoning with change at both metaphysical and qualitative levels because the variable x is structured as an object that does not change over time.[21] While events *occur* and objects *exist*, standard symbolic notation has no way to distinguish between these two modalities.[22] Seeing processes as primary and constitutive of provisional objects changes the status and function of variables. For Whitehead 'there is no such entity as a mere static number' (MT: 93). Rather, numbers are characterisations of the manner and active process by which a 'group unity has been attained' (MT: 93). Like the 'sparrowing' manner of occasions, numbers are more precisely *numberings* than static objects or entities.[23]

Whitehead asks us to 'conceive the fusion of two groups, each characterized by triplicity, into a single group' (MT: 90–1). The characterisation is specific to the aggregates in question – the *how many* question depends upon a choice of what to count and when to count it. In some cases, this counting is the activity that 'makes' the entities. It is therefore 'not true that this process of fusion *necessarily* issues in a group of six *in which the same principle of identifying individual things is preserved*' (MT: 91, emphasis added). As an example, 'consider drops of water, each drop with its own skin of surface tension' (MT: 91). In this case, combining two groups of three drops into a single group 'may result in a coalescence so that one drop results; or it may result in shattering the original drops so that a group of fifty drops appears' (MT: 91). If the arithmetical relation (3+3=6) is analytically true, its application does not hold universally unless we presuppose *a universal principle of individuation* to hold across both sides of the fusion. As in the example of the water drops, this need not necessarily be the case.

This does not mean that anything goes. 'Three plus three equals six' remains an analytically true description of numbers taken as entities in themselves. But Whitehead's formal discussion brings into view a manner in which such an analytical truth *in principle* can tempt metaphysical thought to deny the reality of time, change and context. This foreshadows further shifts discussed in theses 2 and 3, specifically the notion that an individual is relative to a *principle of individuation* that does not function as a static universal. For this reason, it is only generalisable at a level of abstraction that suspends specific determinate content. If any individual, for Whitehead, is a result of the 'satisfaction' of an actual occasion, there is no single universal principle of what such satisfaction requires or entails. Debaise observes: 'satisfaction is not a common end, identifiable with all others, as if there were a pre-existing finality in individuation that would be actualized in a particular manner' (2017a: 78). Because 'there are specific differences between the 'satisfactions' of different entities' (PR: 84) the principle of individuation may manifest differently depending on the 'satisfaction' in question. Whitehead thus challenges the tendency to assume that identity of variables across transitions is formally unproblematic, an assumption as ubiquitous as it is puzzling. Such presumption that a variable necessarily maintains a self-identity even as 'reasoning elaborates novel compositions' (MT: 107) exemplifies what Deleuze calls the dogmatic image of thought's 'moral' choice to privilege the 'Same' as the necessary feature of representation (DR: 130–2).

Deleuze's discussion of the impersonal basis of individuation moves such formal considerations closer to the existential. Deleuze frequently discusses individuation in the context of life and what he calls the impersonal or pre-personal: *a life rather than 'my' life.*[24] The indefinite article does not indicate a generic or generality (see thesis 2), but rather the lack of subjective possession and relative superficiality of ego-identity in the processes constitutive of the individual. Processes of a life are ontologically prior to, even as they are the condition of possibility for, consolidation into a subjective *cogito* or ego. When Deleuze writes that 'Individuation emerges like the act of solving a problem' (DR: 246) the point is not subjective. This would be entirely question-begging. The problem-solving in question (see thesis 3) is ontological, not the wilful act of an agent. Moreover, because individuation occurs in a pre-personal field, identity ascriptions can only be secondary, especially if they are understood as acts of separation or insulation from the pre-personal field that is the continuing source of the individual's emergence. Individuation proceeds according to intensive processes and these cannot be fully determined in the abstract or in general. This is to say that individuation is not *the same for everyone*. The risk of blurring the identity/individual distinction is how it covers over what Deleuze calls individual singularities in favour of a general identity that is conceptualised as static. When Deleuze writes that: 'The individual finds itself attached to a pre-individual half which is not the impersonal within it so much as the reservoir of its singularities' (DR: 246) he refers to the distinction between the extensive and the intensive. The individual is in this case the extensive expression of intensive individuating processes that are inflected through thresholds and singularities of emergence.

Can we encounter this 'pre-individual' or orient towards this reservoir of singularities? How do preconceived notions of identity inhibit attention to differences and events that may function as transforming singularities? Attuning to the dynamism of events depends on the manners of event in question. Nevertheless, because the 'self' is no longer a static substance defined by fixed predicates, this conceptual turn can infiltrate conceptions of self to enable more creative responses:

> Instead of a certain number of predicates being excluded from a thing in virtue of the identity of its concept, each 'thing' opens itself up to the infinity of predicates through which it passes, as it loses its center, its identity as concept or as self. The communication of events replaces the exclusion of predicates. (LS: 174)

This does not erase the individual but does suggest a different attitude towards persevering. To persevere is to pass on a manner or style of inflection, rather than a static identity. Events communicate: such communication is the source of what might function as a lived identity. Rather than attachment to stable identity, the individual is a manner of acting or passing on this communication ('Individuals are signal-sign systems' (DR: 246)) that may change apparent identity in relation to circumstances and events.

All individuation occurs in complex processes of negotiation with constraints that exceed the subjective. But this process of negotiation (1) is not a linear determination and (2) does not end in a discrete individual independent of this process. The *perceived* necessity of rigidly identified constraints may perpetuate conditions that reproduce generic identities that fail to individuate to the fullest intensity possible. In contrast, without fixated attachment on generic identity, individuation emerges in response to requirements and obligations relative to contexts or niches, and this can involve novel, and for Deleuze more intense, achievements of satisfaction that change perceived identities or the relevant features of a context or niche.[25] Though there is no guarantee that such events or achievements will be harmonious, this means being cognisant of the way preconceived identities shape attention – what we notice and do not notice – even as it encourages an experimental stance with regard to the 'given-ness' of that attention and those identities. Dislodging the presumed neatness of fit between individuality and identity opens a space for the emergence of individuals which challenges strictures carried by essentialist identity.

Thesis 2: the singularity/universality relation is not the same as the general/particular

A process approach to individuals reconfigures the relationship between generals and particulars. At issue in individuation is a question of resemblance *and* difference. Generals (forms, essences or generic definitions) explain the resemblance of particulars. In relation to thesis one, identity becomes a general category, even as it refers to a particular: Particular (x) is a General (y). This is easily enlisted into projects of prediction: 'because (x) is a (y), therefore . . . z'.

This schema is constitutive of the traditional problem of individuals. It cannot be simply rejected since it does significant epistemic and metaphysical work. Deleuze indicates awareness of this in declaring 'the task of modern philosophy is to overcome the alternatives . . . particular/universal' (DR: xxi). Whitehead similarly states that 'All modern philosophy hinges around the difficult of describing the world in terms of . . . particular and universal' (PR: 49). Though their efforts to overcome the reductive operation of this opposition are by no means equivalent, they share significant convergences in strategy. Both involve making the transcendental level immanent to, though not identical with, the actual.[26] This complicates hierarchies often generated by Platonic or quasi-Platonic realisms.[27] Both in some sense also universalise the 'general' – that is, the reality of the transcendental applies universally (what Deleuze calls a 'concrete' universal rather than 'an abstract general') even as this 'application' is *not* construed as top-down instantiation or relation of resemblance. This results in a non-traditional philosophical 'realism' (in the medieval sense) that is simultaneously neither a humanism nor a reductive materialism. In Whitehead's words, such efforts hope to rehabilitate 'that immediate experience which we express in our actions, our hopes, our sympathies, [and] our purposes . . . in spite of our lack of phrases for its verbal analysis' (PR: 49).

Thesis 1 emphasised the importance of not blurring identity and individuality because of the extent to which identity categories hinder individuation processes by prematurely closing their outcomes. Thesis 1 was therefore concerned with identity ascription, not the formal principle of identity as such. In this thesis, we turn to look at the principle of self-identity (i.e., x=x), beginning with how Whitehead nuances this principle.

Though he keeps the *principle* of self-identity, Whitehead moves its application from ordinary objects to what he calls 'actual occasions' or 'actual entities'.[28] Actual occasions are the metaphysical atoms of Whitehead's system ('the philosophy of organism . . . is an atomic theory of actuality' (PR: 27)). This affirmation of metaphysical atoms has led Graham Harman to claim an unsurmountable opposition between Deleuze as a 'philosopher of becoming' and Whitehead.[29] However, Whitehead's description of actual occasions complicates the image of 'atom' as a discrete entity materially located in space and time. For one, their relational description means that the 'actual occasion' is less a unit or thing, but rather an *activity* of *unifying*: '*how* an actual entity *becomes* constitutes *what* that actual entity *is* . . . Its "being" is constituted by its "becoming"' (PR: 23). Furthermore, determination of the actual entity is correlative with what Whitehead calls its perishing: 'an actual entity has "perished" when it is complete' (PR: 82). Such determination changes its mode of existence from actual to objective, becoming part of what Whitehead calls 'public matters of fact' (PR: 22).[30] Once the actual occasion achieves satisfaction (through what Whitehead calls 'concrescence'), it perishes, it no longer *is*, in the sense of the immediate actual. And yet this does not mean that it no longer exists. Though debate continues over the status of the actual occasion after it has perished in relation to the genesis of the present actual, on both the traditional and Judith Jones's 'ecstatic' reading, the perished actual occasion still has a mode of existence.[31] To perish is not simply to disappear like an on/off switch, but rather to diminish in its intensity of actuality. Moreover, if actual occasions are always the locus of the most immediate real, they are not exhaustive of all existence, with Whitehead including seven additional categories of existence.[32]

This tension between perishing and remaining relevant as 'data' for subsequent concrescences and actual occasions is deeply embedded in Whitehead's work and is why his metaphysics *is relational through and through*. If from the perspective of the principle of self-identity, each actual occasion just is what it is, its constitution and relevance trouble the exclusivity of this identification: 'the pragmatic use of the actual entity, constituting its static life, lies in the future. The creature perished *and* is immortal' (PR: 82).[33] This sets the context for Whitehead's reworking of the general/particular relation. How does Whitehead explain resemblance between occasions through appeal to 'Eternal Objects' while not assimilating these objects into a model of static general forms?

For readings which stress empiricist motivations and effects in experience, the eternal objects represent a significant challenge, since of necessity they exceed the domain of experience.[34] It is therefore crucial to focus on *how* eternal objects relate to actual occasions. This is particularly important if we are to understand why they are immanent to, rather than transcendent over, experience.

Whitehead presents eternal objects and actual occasions as the two 'fundamental types of entities'. All other types (societies, nexus, forms of relation, and so on) presuppose these two ('the other types of entities only express how all entities of the two fundamental types are in community with each other, in the actual world' (PR: 25)).[35] The distinction between these fundamental entities incorporates two traditional distinctions: (1) between the potential and the actual; and (2) between universals and particulars.[36]

Eternal objects and actual entities map onto the language of universals and particulars: 'the antithetical terms "universals" and "particulars" are the usual words employed to denote respectively entities which nearly, *though not quite*, correspond to the entities here termed "eternal objects," and "actual entities"' (PR: 48, emphasis added). Whitehead challenges formulation of their relation as a unidirectional dependence where the particular is exhaustively characterised through the universal:

> The [traditional] notion of a universal is of that which can enter into the description of many particulars [individuals]; whereas the notion of a particular [individual] is that it is described by universals, and does not itself enter into the description of any other particular [individual]. (PR: 48)

Because it denies the relevance of relations *between* actual entities, Whitehead rejects it: 'an actual entity cannot be described, even inadequately, by universals; because other actual entities do enter into the description of any one actual entity' (PR: 48).[37]

Whitehead's point is not only with regard to description. That is, an actual occasion is metaphysically constituted both through how it 'prehends' (PR: 18–20) previous actual occasions *and* how it realises relevant potentials (eternal objects). The very 'being' of any actual occasion, is related to all other actual occasions, in *gradations of relevance* or, to use Judith Jones's phrase, *intensity*.[38] These relations range from the largely negligible (what Whitehead calls a 'negative prehension') to the nearly determinate. Every actual occasion is to some extent constituted by its relation to all other occasions: 'if we allow for degrees of relevance, and for negligible relevance, we must say that every actual entity is present in every other actual entity' (PR: 50).[39] Eternal objects thus do not determine actual occasions full-stop because the 'horizontal' relations between occasions are equally relevant to how the eternal object is realised in the contemporary occasion.

The mode of realisation between potential eternal object and actual occasion also complicates top-down instantiation models in denying a one-to-one resemblance. Relations between eternal objects and actual occasions proceed through what Whitehead calls 'ingression' (PR: 29). Ingression expresses the transition from the *potentiality* of the eternal object into actuality as a form of definiteness. From the perspective of the actual occasions, the 'same' eternal object can ingress into occasions in different manners and degrees (PR: 29, 39–41). This is why Whitehead states that an eternal object 'involve[s] *indetermination* in a sense more complete than' actual occasions (PR: 29, emphasis added). While eternal objects are 'forms', they are, as Deleuze and Guattari say 'anexact but rigorous' (ATP: 367).[40]

Though Whitehead correlates 'eternal objects' with 'Platonic forms' (PR: 44), this includes the caveat that the form is not a discrete object, but rather a relational multiplicity. As relational multiplicity, ingression into actual occasions occurs in varying ways.[41] In a sense, Whitehead's metaphysics doubles relationality. Eternal objects represent, as Elisabeth Krauss puts it, 'the abstract patterns' that inform the 'patterned interfusion of . . . events' that are actual occasions (1998: 31). Neither actual occasion nor eternal object can exist as a fully determined and isolated discrete entity *in actuality*. For actual occasions, this is because their full determination is the event of their becoming no longer fully actual. For eternal objects, this is because they are essentially relational potentials ('forms of definiteness' (PR: 22)). We can think of them as a complex of relation in which their ingression is linked to other ingressions. 'If this happens, that happens, which shifts the levels of all of the other objects.' Eternal object *qua* eternal object is more abstract because it is not fully determined *until* ingression and the intensity of this ingression varies: '[eternal objects'] conceptual recognition does not involve a *necessary* reference to any definite actual entity of the temporal world' (PR: 44, emphasis added).

How the eternal object ingresses into the actual occasion depends on the actual occasion, not vice versa. *Qua* eternal object in-itself, 'it is neutral as to the fact of its physical ingression in any particular actual entity of the temporal world' (PR: 44). The reason for ingression is not found in the eternal object alone. Potentiality does not determine actuality; it is actuality that determines potentiality. This is encapsulated in what Whitehead refers to as the 'ontological principle': 'no actual entity, then no reason . . . [or] actual entities are the only *reasons*; so that to search for a *reason* is to search for one or more actual entities' (PR: 19, 24)).[42] This ensures that Whitehead's rational construction of eternal objects does not seek to legislate the empirical from a standpoint of pure rationality. The eternal objects are eternal in their potentiality, but because they are neutral as to their manner and degree of ingression we cannot use them to necessarily predict future actual occasions. In this sense, they depart very much from Kantian categories. As Jean Wahl writes: 'Eternal objects [. . .] tell us nothing about their ingression in experience. In order to see them, there is only one thing to do: adventure in the domain of experience' (2004: 135).

Whitehead therefore denies a relation of top-down determination often ascribed to the general/particular relationship. The eternal object is not a set that groups together all particulars sharing a certain quality. Rather, each actual occasion determines the potentiality of all relevant eternal objects to manifest. Whitehead refers to this as 'decision': the decision of the actual occasion 'cuts off' or excludes potential degrees of the eternal object into a precise actualisation. Such decision is necessary for the achievement of any concrete actuality. The process of concrescence is impossible without eternal objects as potentialities, but this process is a form of limiting pure potentiality in a determinate realisation: "'Potentiality' is the correlative of 'givenness'. The meaning of 'givenness is that what *is* "given" might not have been "given"; and that what *is not* "given" *might have been* "given"'" (PR: 44). Because the eternal object *qua* eternal object exceeds any particular manifestation in an actual occasion, this destabilises any claim to have exhausted experience with a necessary certainty. We cannot presume that potentialities are fully understood or mapped out in advance. The eternal object is not equivalent to its manifestation in experience.

At this point it is helpful to recall why eternal objects were posited at all. Recall, eternal objects serve as conceptual construction necessary for explaining how it is possible that occasions resemble one another. Their *eternal* quality as such names fixed potentials of relations as the condition for what Whitehead calls the 'solidarity' of the universe (PR: 40). Eternal objects refer to a potentiality that is everywhere, for all time and place, even as the manifestation of this potentiality remains contingent to the perspective of the actual occasion (both through its subjective aim and decision). The excess of the eternal object to particular occasions is crucial for a metaphysics capable of explaining genuine change. Without this excess, 'the alternative is a static monistic universe' (PR: 46). Though eternal objects partially explain relative permanence in conditioning resemblances between actual occasions, their more consequential function is as reservoir for change. This is because their conditioning of the actual occasion is *not* a top-down determination nor a direct representation of the object in question. They are not 'general' categories.

Whitehead sometimes uses colour as a way of making eternal objects more intuitive. However, colour is not a separate static quality, but rather a qualification of the event in which it appears. It is not subject (balloon) with predicate (red), but event characterised by differing manifestation of eternal objects ('circl-ing red-ing here'). Krauss describes this in terms of eternal object as 'adverb, rather than adjective': 'Just as the eye does not see a red object but appropriates a certain environmental event "redly," so the hand might appropriate the same event "hotly"' (Krauss 1998: 31) This is in keeping with Whitehead's enigmatic claim that 'We experience the green foliage of the spring greenly' (AI: 250). Nevertheless, something about these events refers to eternal potentials. It is both *precise* to the occasion in question – that gorgeous spot of blue there – and eternal: blueness as a form of qualification that is yet excessive to any particular manifestation and can, moreover, be investigated. Blue under new conditions of temperature, pressure, light; blue as seen through different eyes.

This means that 'blue' *as such* is not equivalent to 'eternal object'. Rather, it names a gesture towards a relational potential that does not carry with it a single principle of individuation. The manner of inflecting an actual occasion will appear in *differing degrees of intensity*. From the perspective of the actual occasion, this ingression is precise, not vague. But no single eternal object determines the actual occasion full-stop, rather the occasion's decisions involve 'selection from these forms [eternal objects] . . . grade[d] *in a diversity of relevance*' (PR: 43, emphasis added). The decision of actuality inflects all forms of potentiality into a single occasion.[43] The actual occasion achieves togetherness, *a feeling together*, that mediates between potentialities (eternal objects) and actualities (past occasions that are now objective datum). Whitehead's philosophy of organism therefore recognises historical and material factors as binding the innovations possible at any particular actual occasion, while likewise opening a space such that *unexpressed potentialities* may emerge in the future.

Individuals escape the assumptions of a general/particular format because the relationship between the two categories is not one of resemblance and the movement from one to the other is not accomplished through a mode of representation. What an individual actual occasion *is* cannot be given in advance because it neither resembles nor instantiates a universal essence. Since each actual occasion, as a unique concrescence, is nevertheless in some relation, however negligible, to all other actual occasions, this makes it a kind of universal – *a singularisation of universality*. It draws the universe together into a new moment of becoming, a new achievement or drop of subjectivity, as Kraus puts it.[44] This achievement is not equivalent to the ordinary mode of lived (human) subjectivity, but is rather, in some sense, the condition of possibility for this lived subjectivity. Whitehead's 'transcendental' posit of eternal objects provides a way of accounting for the singular identity of the occasion while nevertheless also enabling some durational persistence. But this durational persistence is not a form of strict identity. It is not a particular that instantiates a universal. Rather, if the actual occasion just is what it is ('Actual occasions perish, but do not change; they are what they are' (PR: 35)), at the level of durational experience, there is no static identity. Actual occasions singularise, but their grouping into relative durational persistence is a matter of, as Debaise puts it, 'instauration', in which what is renewed is a manner of feeling that is not identical but is rather responsive to the dual constraints of the inherited feelings from posterior occasions and the eternal object (2017b: 67).

Deleuze's 'virtual' has a similar function as Whitehead's eternal objects in developing a realist philosophy that resists transcendental idealism *and* static essentialist Platonism. Like eternal objects, Deleuze reconfigures the relation between potential and actual such that the conditioned real does not *resemble* the conditioning virtual. However, there is a significant divergence in technical emphasis with regard to the principle of self-identity. Deleuze has nothing that is as determinate and comprehensive in scope as the actual occasion. Recall, following the 'ontological principle', actual occasions 'are the final real things of which the world is made up. There is no going behind actual entities to find anything more real' (PR: 18). But this is also to say that the actual occasion, 'caused' by other actual occasions and informed by eternal objects, *as a category* cannot track differences in consequence. While 'there are gradations of importance, and diversities of function . . . *in the principles which actuality exemplifies all are on the same level*' (PR: 18, emphasis added). The actual occasion is the individual, but at the abstract level of occasions *qua* occasion, they are to some extent *'all the same' in their difference*.

In keeping with the overriding stricture of Deleuze's thought to not presuppose identity or the 'same', his thought does not posit anything as precisely individual as the actual occasion. If for Whitehead the actual occasion is the only technical individual, for Deleuze, processes of individuation always proceed from a pre-individual field.[45] This field is transcendental in providing the conditions for real experience. However, while it is true that it cannot be represented, it remains, in a sense, lived or felt, as we have seen in Chapter 1. This is why his philosophy remains an empiricism: we do not know the limits ahead of time and they are not to be discovered through purely rational or conceptual means. What is the character of this transcendental or pre-individual field? Deleuze insists that 'We cannot think of the condition in the image of the conditioned. The task of a philosophy which does not wish to fall into the traps of a consciousness and the cogito is *to purge the transcendental field of all resemblance*' (LS: 125). If we do not do this, and 'retain the form of the person, of personal consciousness, and of subjective identity' then we are failing to explain how the sense of an individuated person is produced. We are simply transposing it to the transcendental level: 'everything which must be engendered *by* the notion of sense is given *in* the notion of sense' (LS: 98).

However, while it is true that Deleuze does not want to characterise the transcendental through categories of representation, Deleuze does describe some characteristics. Most notably the transcendental is intensive and continuous, rather than discrete or extensive. Extensive forms and qualities are in a sense produced by intensive differences at the transcendental level. Here then appears a divergence with Whitehead regarding their respective characterisations of the transcendental: do we have actual *discontinuity on the basis of transcendental continuity (Deleuze), or actual continuity (Whitehead) as produced on the basis of transcendental discontinuity?*

Deleuze endorses an intensive continuity of differential becoming such that any extensive identity emerges out of a more primary play of intensive differences, whereas Whitehead endorses a metaphysical atomicity that relies upon apparently quasi-discrete eternal objects and actual occasions. While both understand macro-level individuals as secondary to events, patterns and processes, they explain the 'ground' of these processes differently. Whitehead sees continuity as an outcome of more primary processes of becoming that are atomic:

> the extensive continuity of the physical universe has usually been construed to mean that there is a continuity of becoming. But if we admit that 'something becomes,' it is easy, by employing Zeno's method, to prove that there can be no continuity of becoming. There is a becoming of continuity, but no continuity of becoming. The actual occasions are the creatures which become, and they constitute a continuously extensive world. In other words, extensiveness becomes, but 'becoming' is not itself extensive. (PR: 35)

Deleuze reverses this order of priority. Understanding the metaphysical status of what he calls 'problems' or 'Ideas' requires 'a Riemannian-type differential geometry which tends to give rise to discontinuity on the basis of continuity, or to ground solutions in the conditions of the problems' (DR: 162/210). For this reason, 'To solve a problem is always to give rise to discontinuities *on the basis of a continuity which functions as Idea*' (DR: 162, emphasis added).

Does this apparent divergence make an existential difference? One is tempted to downplay it because in either case lived experience, at least in ordinary modalities, occurs in a manifold of continuous extension. But it is also the case that the divergence may not be as formally significant as it first appears. Deleuze is concerned to show that the solutions to a problem do not exhaust its virtual or transcendental potential. The problem, in a sense, subsists beneath differing solutions. The solutions are discontinuous in that they determine and actualise the problem, but they do not exhaust it. The problem remains, percolating, if you will. We do not experience 'the' problem as such, but experience is a manifestation of problematic intensities as its transcendental conditions. Actualising or singularising these conditions into an individual experience means selecting or disrupting their continuity. It is, if you will, the 'pulse' of the actual occasion. So Deleuze and Whitehead are discussing two different senses of continuity. Whitehead's becoming of continuity is a becoming of extension understood as the extensive field in which measurement occurs in empirical experience. Deleuze's becoming of discontinuity is referring to the way that the 'solution' or empirical manifestation of an intensive or virtual problem singularises the problem without annulling it. What else is such a solution but the actual occasion and its manner of concrescence from the forms of potential (eternal objects)?

However, a certain difference remains consequential. Where Whitehead universalises singularity in the category of the actual occasion, Deleuze instead emphasises certain relational differences that make a difference – which he names as singularities. Singularities are primarily transcendental events ('ideal events') that mark moments of consequential change in intensity or trajectory. Such ideal events 'are more profound than and different in nature from the real events which they determine in the order of solutions' (DR: 163). A singularity differs both from actual occasions and eternal objects insofar as it includes an evaluation of importance or relevance. However, because a singularity is ideal or transcendental, not actual, it functions, like the eternal object, as immanent potential rather than transcendental form.

Singularities occur in contexts or fields. Deleuze calls these fields a 'problem' or an 'Idea'.[46] Problem-ideas are continuous intensive multiplicities that are virtual rather than actual: 'Problematic ideas are not simple essences, but *multiplicities or complexes of relations and corresponding singularities*' (DR: 163, emphasis added). Singularities thus are relative to problem-ideas insofar as they distinguish between ordinary events and those that transform the nature of potential resolution to the problem. As Smith observes, because 'Every multiplicity is characterized by a combination of singular and ordinary points . . . One could say that there are 2 poles of Deleuze's philosophy: "Everything is ordinary" and "everything is singular"' (Smith 2012: 115). This distinction between ordinary and singular is not found in Whitehead's scheme at the transcendental level.

The second wrinkle involves the characterisation of virtual multiplicities as 'problems'. Because problems and singularities inform transcendental complexes or multiplicities, they bear a significant relation to Whitehead's eternal objects. While we might be tempted to thus see their terminological differences as primarily verbal, the choice to stress problem (Deleuze) or eternity (Whitehead) engenders differing degrees of affective intensity and drama. On the one hand, there is an increased sense of active precarity and struggle (problem), on the other, a gesture towards the 'peace which passeth all understanding'. This point can be overstated though. 'Problems' are not primarily subjective conditions, rather, a problem is a metaphysical 'objecticity' (and in this sense perspective on 'struggle' may be transformed if one takes as the goal of struggle the cessation of problems – an impossibility for Deleuze). Problems are 'ideal' in being both virtually real but not exhausted by any particular actualised solution. They do not refer to a condition of lack or flaw, but instead are 'ideal "objecticities" possessing their own sufficiency and implying acts of constitution and investment in their respective symbolic fields' (DR: 159/206). They are virtual differential conditions which give rise to different determinations of actualised problem-solutions. Such problem-solutions are relative to a field of individuation that may be psychic, social, biological, chemical, physical, and so on.

Problems replace general categories and singularities replace the specific differences that inflect the thresholds between individualised or particularised actualisations. Such actualisations can be thought of as 'solutions' to these problems, provided we observe some caveats. Even if, as Daniel Smith observes 'Being always presents itself under a problematic form' (2012: 136), this does not mean we can isolate the fundamental problem or the problem of all problems. To characterise a problem as a general concept is both too broad and too narrow. It is too broad because it cannot characterise the real differences within its particulars, and it is too narrow because it presumes to denote against a neutral or homogenous background. By contrast, when the presupposition *is* difference (rather than homogeneity), then Deleuze declares that 'the problem . . . is a concrete singularity no less than a true universal' (DR: 163). The concrete singularity refers to the problem insofar as it is immanent to an actualised solution, the true universal insofar as this solution does not exhaust the potentiality of the transcendental problem ('the problem is at once both transcendent and in relation to its solutions' (DR: 163)). There is no universally homogeneous problem *as such* even as all reality is constituted by complexes of problem-solution. At the same time, a problem is true universally, even as it is actualised in different possible resolutions according to changing conditions and contexts.

Second, no particular actualisation or resolution of a problem exhausts its transcendental potentiality. The problem does not exist as an essential form that is *the same* in every case, since it involves complex loops of feedback with other problems as well as changing conditions of actualisation. It is here that the 'problematic distinction between the ordinary and the singular' becomes crucial (DR: 163). Using a graphical model, consider a differential curve that integrates a number of different relational values. Each problem is composed of sub-problems and problems are composed of relations between differences. Thus, at certain points of inflection on this curve (within this problem), dramatic changes will occur. These are its distinctive points ('Corresponding to the relations which constitute the universality of the problem is the distribution of singular points and distinctive points which determine the conditions of the problem' (DR: 163)). At other points, the curve continues its ordinary pattern.

To make this intuitive, James Williams adopts the rhetorical strategy of expressing problems in terms of infinitive verbs (keeping in mind that problems are not linked to human experience only). Consider the differential problems expressed in the following infinitives: to fall in love, to find God, to give birth, to live well.[47] For each, moments of ordinary habit interact with event-singularities where the problem passes a threshold. Falling in love means one thing, is lived one way, until the event-singularity that changes its manifestation.

We must take care with existential examples however, precisely because the event-singularity is ontologically separate from the living individual that lives through it (and is thus inflected and altered). The event-singularity is a 'transcendental event' insofar as it belongs to the transcendental field purged of all resemblance to the lived ego and its representation. But, the event-singularity is double-sided, as the example shows, insofar as it is actualised in existential terms. This actualisation does not simply instantiate nor resemble nor exhaust its virtual condition. From the virtual perspective, event-singularities are 'anonymous . . . nomadic . . . impersonal . . . pre-individual' (LS: 103). Whatever the event which actualises a shift in the problem (to fall in love) – it does not belong to the living subject alone, nor is its actualisation confined to that subject. As events of the impersonal transcendental field, event-singularities express points of significance 'wherein the differences between series are distributed' (LS: 103). Their relational complexity is metaphysically prior to the borders or separated stability of a macro-level Self, Subject, object, or entity:

> Far *from being* individual or personal, singularities *preside over* the genesis of individuals and persons; they are distributed in a 'potential' which admits neither Self nor I, but which produces them by actualizing or realizing itself, although the figures of this actualization do not at all *resemble* the realized potential. (LS: 103, emphasis added)

Singularities determine individuals, not vice versa.

As virtual, they are not exhausted in actualisations and they subsist whether or not they are actualised. Even if an actual individual does not actualise the 'boiling point' singularity, for example, this singularity is still real in marking a potential shift or place where a curve or series (that is, a mapping of relations between changes or differences) changes character. As realm of differential relations, the virtual is akin to potentiality. Potentiality however, following Bergson, is *not* possibility. Possibility works only at the level of the extensive or what Deleuze calls the 'proposition' (DR: 162) by tracing its conditions off of the actual. Possibility in a sense compares differing states of affairs. It is this way, but it could have been this way. The house is painted yellow, but it could have been painted red. It is in this sense entirely 'sterile' to use Deleuze's expression in *Logic of Sense*. Potentiality, by contrast, is ontologically real without being actual. Potentiality operates within the intensive and pre-individual fields that give rise to extension but are not exhausted by these actualisations.

Examples from the physical sciences can also be used to concretise this concept of a virtual singularity. Recall, no single symbolic field exhausts a transcendental problem even as its actualisation is not one of resemblance. The 'boiling point' of the subjective individual is not the 'same' as the boiling point of water, for example, and yet there is a resonance. DeLanda uses the example of a 'phase transition':

> phase transitions are events which take place at critical values of some parameter (temperature, for example) switching a physical system from one state to another, like the critical points of temperature at which water changes from ice to liquid, or from liquid to steam. (2002: 10–11)[48]

A boiling point is an ideal event: a singularity that is actualised in the event of boiling. Such actualisation is unique and precise but also differentially related to other conditions (altitude, air pressure, etc.) that are themselves also understandable as differentially related. In this way, there is *both* recognition of the singularity of threshold points that exist as virtual potentials informing the emergence of actuality, as well as an insistence on the relational influence of other series of differential relations. The differential relations are virtual transcendental multiplicities that replace static or collective conceptions of generality. The singularity marks the event of a multiplicity undergoing a qualitative transformation. It is a threshold whereby qualitative experience undergoes such a change that the identity appears to change (what Deleuze calls a 'becoming').

Actualisation and individuation involve *actual* events that are partial and relative to a pre-individual context or field. This process involves a transition from the intensive realm of pre-individual forces to the differentiated extensive realm of discrete objects. Because this process never exhausts the metaphysical problem, there is a significant resonance with the eternal part of eternal objects. James Williams observes: 'A problem for Deleuze is *never* resolved. Instead, it interacts with different times in different ways such that each must find the best way to balance the positive and negative effects by transforming it ...' (2010: 26, emphasis added). This gives a way of thinking an individual event as a resolution to a problem that cuts across categorical distinctions. Such events may be those which befall a human life or those which occur at the juncture of various physical fluids. Aden Evens gives the example of rain dripping onto a leaf: 'the paths it takes across that uneven surface are solutions to a complex problem at the juncture of differential topology, fluid dynamics, evolutionary biology, and metaphysics' (2000: 106). Each actualisation is a solution/ event based on the differential conditions in a field of individuation.

Distinguishing between ordinary and singular points is integral to Deleuze's development of a realist philosophy that is dynamic at the transcendental level. Moreover, by denying a relation of resemblance between transcendental conditions (eternal objects or virtual problem-singularity complexes), both Deleuze and Whitehead affirm the irreplaceability of attention to ongoing events of experience. Events manifest potentials that cannot be necessarily predetermined from any single rationalist perspective, even as these potentials are to some extent determinative of the limits of the perspectives they engender. This metaphysical shift entails a different prioritising of attention with regard to what is individual. Because individuating is a dynamic process which does not instantiate a fixed general category, we cannot presume that such categories (whether empirical or transcendental) identify stable natural kinds. For an ecological attunement, we will have to orient attention under the principle that there is no universal principle of individuation that functions identically across disparate domains. This is significant in disrupting the tendency to pay attention only to a single scale of sense or causality.

Thesis 3: individuations require at least two disparate fields or series and are inherently relational

For both Whitehead and Deleuze, processes of individuation are inherently relational. Another way of putting this: individuation emerges as an effect of encounters *between* differences. An individual exists as a node or juncture of informational transfer even as such transfer exceeds intentionality. But, if an individual is for this reason an effect of processes metaphysically prior to their emergence, they are also participant in such processes at different levels. An individual is thus constituted by relational encounters, and in this sense dependent upon them, but also makes possible further encounters between differences that would not be actualised without the individual in question.

The most important influence for this aspect of Deleuze's approach to individuation is the work of Gilbert Simondon. Deleuze declares that Simondon's *L'individu et sa genèse physico-biologique* 'presents the first thought-out theory of impersonal and pre-individual singularities' and as such offers a 'new conception of the transcendental' (LS: 344, fn3).[49] These innovations occur in the context of Simondon's efforts to think individuation in a manner different from what he identifies as the two paths of previous Western philosophy: the 'substantialist' or 'hylomorphic' paths (1989: 9). Neither path offers a manner of thinking what Simondon calls 'ontogenesis' (1989: 10). The former, as we have already seen, presumes the individual as a given fully constituted and independent entity. The latter operates instead according to a 'bipolarity' of form and matter, but still presumes the form as already individuated.[50] In both cases, whatever principle of individuation is posited is a 'principle of reflection' only, not a truly 'genetic' principle (DI: 86).[51]

Simondon instead proposes that we begin with 'preindividual being' (1989). His innovation is to take a concept from thermodynamic sciences for helping to think this 'problematic' state of being. (This is why Deleuze credits Simondon for 'demonstrat[ing] the extent to which a philosopher can both find his inspiration in contemporary science and at the same time connect with the major problems of classical philosophy – even as he transforms and renews those problems' (DI: 89).) The concept is 'metastability': in physics, metastability refers to a stable state of a dynamic system that is not a state of least energy. That is, the state is relatively stable but has not exhausted its potentiality for change. Its equilibrium is therefore 'metastable'. In this way a system is metastable when 'the least modification of system parameters suffices to break its equilibrium' (Combes 2013: 3).

For Deleuze following Simondon, this condition of metastability offers an abstract manner for thinking being as a 'pre-individual' field. Such a field is in excess of itself: 'Preindividual being . . . harbors potentials that are incompatible because they belong to hetereogeneous dimensions of being' (Combes 2013: 4). These potentials are not isolated objects but rather involve what Deleuze calls 'disparateness' – potential is always a measure of encounter between differences. DeLanda offers an example from information theory as one way of conceptualising this: 'When two separate [potential] series of events are placed in communication, in such a way that a change in probabilities in one series affects the probability distribution of the other, we have an information channel' (2002: 79). A pre-individual field can thus be characterised as containing potential or virtual events in relation to each other. As Deleuze writes drawing on Simondon 'individuation presupposes a prior metastable state – in other words, the existence of "disparateness" such as at least two orders of magnitude or two scales of heterogeneous reality between which potentials are distributed' (DR: 246).

From this metastable state, such a pre-individual field can be tipped out of its equilibrium by alterations or triggers (these are what Deleuze refers to variously as the 'quasi-casual operator' (LS) or 'dark precursor' (DR)). The causality here cannot be linear or efficient. If it were, then the whole transition of a (virtual or potential) pre-individual to an actualised individual would be question-begging. Indeed, such a trigger or quasi-cause itself must be construed in terms of communication between disparates. As Deleuze writes: 'Simondon's conception . . . can . . . be assimilated to a theory of intensive quanta . . . [where] each intensive quanta in itself is difference' (DI: 87).[52] In a sense then, the dark precursor or quasi-casual operator names what must be the case for individuating structurally, but it cannot be identified or represented.

Be that as it may, the relevant implication is that the encounter between two disparate scales or series of events does not result in a static outcome, but rather an 'information channel'. This is why Deleuze says that 'every phenomenon flashes in a signal-sign system' (DR: 222). Following Simondon 'the emergence of an individual . . . should be conceived in terms of resolution of a tension between potentials belonging to previously separated orders of magnitude' (Combes 2013: 4). An individual is an expression of differences put into communication. Put in existential language: 'you' are an expression of signal-signs at various levels – the biological, the chemical, the psychic and affective, the semiotic, and so on, with the interior/exterior difference being most vitally implicated. Simondon writes: 'The psyche is made of successive individualizations enabling being to resolve problematic states corresponding to a permanent setting in communication of greater and smaller' (1989: 22).

Simondon is most interested in the way that this individuating process between levels scales upward as well. Just as 'you' are an expression of the signal-signs flashing between levels, you also function as a signal-sign in further networks and collectivities. You flash signal-signs interpersonally, socially, digitally, materially and for this reason: 'the psyche is the foundation of participation in a vaster individuation, that of the collective' (1989: 22). You are then an expression of connective relations between forces at levels that are both intra-individual and trans-individual: potentials of the air, soil, minerals, but also potentials of collective affect, ideological formations and social movement.[53] The emergence of an individual as actualisation from a pre-individual excess of potentiality also informs new series as conditions for subsequent potential metastable systems and processes of individuation. 'As pre-individual, being is more than one – metastable, superposed ... As individuated, it is still multiple, because it is "multiphased," a phase of becoming that will lead to a new process' (DI: 89). Any individual presupposes relations between differences and puts into contact different scales of being such that no single scale can be privileged as dominant or exclusive locus of the real:

> An 'objective' problematic field appears, determined by the distance between two heterogeneous orders. Individuation emerges like the act of solving such a problem, or – what amounts to the same thing, – like the actualization of a potential and the establishing of communication between disparates. (DR: 246)

For Deleuze processes of individuation both depend on an encounter between differences and create *new* differences. On the surface, this is in contrast to Whitehead's positing of the actual occasion as metaphysical atom when atom is conflated with a qualitatively homogenous element. But this divergence is not as stark as it appears for two reasons. First (explored at length in Chapter 4) Whitehead's account of 'societies' – as the nexus of actual occasions generating perduring objects, organisms, and so on – displays scale relativity. There is no one scale of the society as such, but rather societies emerge and interpenetrate on several different scales (a cell is a society, but so is the organism it participates within and so is the social collective in which that organism participates). For this reason, while it is true that Whitehead's actual occasion retains a more rigorous principle of identity than anything in Deleuze, lived individuality is always at the 'social' level and this level necessarily is constituted through its manner of achieving patterns of relational consistency. This achievement is never fixed but is rather perpetually renewed and projected in relation to 'environmental' conditions (see Chapter 4).

Second, while debate remains over the scale or 'size' of the actual occasion, there is no doubt that the occasion is constituted by 'concrescence' of two different orders: the 'prehended' datum of the past and the order of eternal objects or potentialities. The outcome of this concrescence is additionally governed by the 'subjective' or private aim of the occasion (to be explored in Chapter 3).[54] This is to say that the actual occasion itself, while treated as a rigorous individual once it has 'perished', is not simply a homogenous atom. Indeed, the bulk of *Process and Reality* is occupied with internal analysis of the complex processes constitutive of the occasion. These include both positive inclusion of the 'datum' of the past, as well as negative prehension or omission. These positive/negative prehensions are themselves part of the occasion's bipolarity – its emergence as an outcome mediating conceptual and physical feeling. Finally, the ingression of 'eternal objects' into these prehensions, in gradations of relevance and intensity, is yet another way that the unity of the occasion is an outcome of mediation between different levels.

Both accounts thus understand individuation through processes that are inextricably relational. No individual exists as an isolated independent entity; rather individuals are emergent as resolutions between differing series or fields. These fields can be characterised in different ways. Temporally, individuation is always emergent at the intersection between 'past-present-future'. The individual is not given as the stability that persists through time, rather the individual's apparent stability is always being produced through the temporal synthesis of the present. Another way of describing this is in terms of encounter between 'disjunction-unification-addition'. As Whitehead puts it, the many (disjunctive diversity) become one (concrescence of the actual occasion) and are increased by one (passage to objective datum – Nature as passage). Disjunctive diversity is, in a sense, a condition of the past for the present, unification is the process of the actual occasion which constitutes the becoming of the present, and addition is the passage of this present. It is addition that constitutes the movement conventionally associated with time.

These features suggest implications for ecological attunement. Scale variability means that ecological attunement must cultivate awareness of multiple causalities intersecting through any moment, without selecting a single causality as the only arbiter of value or meaning. Relationality combined with scale variability also means that actual occasions can be ingredients simultaneously in different 'individuals'. To be an individual is not to be separate from the possibility that this individuality is constitutive of other individuals as well. Multiple individual actual occasions participate in one another.

If all individuation requires encounters between differing fields, is there anything more that can be discerned regarding these processes? Why do some encounters seemingly result in a coherent individual and some offer no more than 'noise'? Understanding this requires further scrutiny into the operative roles of feeling and affect in processes of individuation. In both cases, feeling and affect are no longer easily assimilated into the language of emotion or states *possessed* by a subject – they are rather metaphysical operations that exceed the boundaries of the subject. For Whitehead then, how are we to understand what he calls the 'subjective aim' of an actual occasion, an aim governed by 'self-enjoyment'? (MT: 150). For Deleuze, given the proliferation of differences both as pre-individual fields and singularised events, how are we to connect this to ordinary modes of lived experience? If individuation no longer refers to 'a person, subject, thing, or substance . . . [but rather] consist[s] entirely of relations of movement and rest between molecules or particles, capacities to affect or be affected' (ATP: 261), how does the living person connect to this process – whether conceptually or existentially?

Notes

1. James struggles to reconcile the speculative direction of his radical empiricism with earlier work in *Principles* around this issue. This is indicated in the fascinating document *The Miller-Bode Objections* (1905–8) which show James's thinking through objections to his radical empiricism essays. A central theme is how to reconcile the possibility of a unit of experience being 'taken' in different ways: 'Can the pen be the "same" (no difference) if it figures in two minds?' (James 1988: 69). Though Deleuze could not have had access to these texts, James experiments with using a concept of the virtual as a means of resolving the issue. Though the experience is actually different, it is virtually 'the same' insofar as it will retroactively be recognised as the same. James is not satisfied with this, because it only works for successive experience, not immediate experience as such.
2. Whitehead makes this point in experiential terms also: 'Being tackled at Rugby, there is the Real. Nobody who hasn't been knocked down has the slightest notion of what the Real is' (quoted in Hocking 1963: 15).
3. See, for example, Hazlett 2010; Kripke 1980; Lowe 2003.
4. Intension refers to the definition entailed by individual. Extension, status and principle are all metaphysical, having to do with, respectively, what *is* individual and the cause or foundation of this individuality. Discernibility is epistemological and linguistic reference is semantic. Gracia mentions process metaphysics in a single footnote referring to its fondness for approaching individuality through the language of 'modes' (1988: 263, n25). He eventually endorses a modal conception of individuality himself, but without engaging process thought in any detail.
5. Though individuality is not the same as mereology (parts and wholes), they are practically related. In most practical senses, it seems to be an individual requires some individuating boundary correlative with being some sort of whole. These concerns are most relevant to issues in philosophy of biology, where they show up both at the level of the organism and the species. Michael Ghiselin for example argues that the relation between individual species and individual organisms should be construed as one of wholes and parts. This view remains controversial. Alberto Toscano shows how the problem of 'life' for a post-Kantian philosophy is understanding the relations between parts of an organism and the organism as a whole – that is, where to place the locus of its individuality. This question follows Kant's recognition that a thoroughgoing Newtonian physics must account for the apparent escape from reductive efficient causality in understanding an organism as purposive in a manner that mere material is not. See DeLanda 2002: 49–55; Ghiselin 1997; Toscano 2006: 19–44.
6. Interestingly, Gracia's eventual conclusion muddies this in appealing to a logic of included middle: 'Universals *qua* universals do not exist or not-exist. As long as this is kept in mind, most of the puzzles associated with the extension of universals and individuals disappear' (1988: 112).
7. The Whitehead quote continues, philosophy 'has been haunted by the unfortunate notion that its method is dogmatically to indicate premises which are severally clear, distinct, and certain; and to erect upon those premises a deductive system of thought' (PR: 8).
8. In this regard, Whitehead comments on what was then Gödel's recent 'Incompleteness theorem': 'Today, even Logic itself is struggling with the discovery embodied in a

formal proof, that every finite set of premises must indicate notions that are excluded from its direct purview' (MT: 2). Gödel is clearly who he has in mind. What has come to be known as the 'Incompleteness theorem' (actually a conjunction of Theorem VI and Theorem XI in Gödel's original paper 'On Formally Undecidable Propositions of *Principia Mathematica* and Related Systems I') is frequently described as showing that there can be no formal system that is both sound and complete – that is, that proves only true statements and all the true statements. There will, of necessity following Gödel's proof, be statements that are undecidable within the context of the formal system in question. Whitehead would clearly know about this result since it originates in the context of his work with Russell in the foundation of mathematics. For further engagement with its philosophical consequences, see Berto 2009. Similarly for Deleuze: 'Logic is reductionist not accidentally but essentially and necessarily' (WIP: 135). Formal logic must select stable propositions to serve as premises. But such selection is a form of leaving out. Since every premise is a selection, this implies some measure of inference that is not being brought into the explicit process of evaluation.

9. This is because these laws are most cogent under a substance metaphysical model, where discrete objects are taken as the paradigmatic existent. See Massumi 2014: 30–6; Smith 2012: 43–51; Zourabichvili 2012: 170–1.

10. For further study of formal possibilities in non-classical logics see Priest 1995, 2008. For insightful work on the connection between these formal results and ontological and political questions, see Livingston 2012.

11. Whitehead's bipolarity of the actual occasion plays a similar function to Deleuze's 'double-causality' between bodies and sense. See *The Logic of Sense*, especially Series 14 through 17.

12. Deleuze and Guattari write: 'The problem of consistency concerns the manner in which the components of a territorial assemblage hold together' (ATP: 327). Speaking most strictly, this denies any discrete individual and instead suggests that an individual is an assemblage.

13. 'This personality extends it*self* beyond present Existence to what is past, only by consciousness, whereby it becomes concerned and accountable, owns and imputes to it self past Actions, just upon the same ground, and for the same reason, that it does the present' (Locke 2008 [1687]: 346). For Plato, I have in mind the theory of recollection and the discussion of reincarnation in the *Phaedo*, to name just two of many examples.

14. Whereas 'individual' could apply to a ball or desk, terms such as self, subject and subjectivity refer to a particular kind of individuality that includes intentional and auto-referential or reflexive experience. Lorraine Code notes that 'self' tends to be the term of choice for discussions in the Anglo-American liberal tradition, whereas 'subject' is more common in the Cartesian and French phenomenological traditions (2006: 201–3). For my purposes, nothing significant hangs on this terminology, so, while I primarily use 'subject' or 'subjectivity' to refer to intentional and reflexive experience of one's 'self' as a living person, these terms are largely synonymous: to be a subject is to have a self.

15. Deleuze talks about 'subjectification' more frequently in collaborative work with Guattari, who additionally makes a distinction between the individual and subjectivity: 'Vectors of subjectification do not necessarily pass through the individual, which . . . appears to be something like a "terminal" for processes that involve human groups, socio-economic ensembles, data-processing machines, etc.' (2008: 25).

16. It is conceivable for an individual to exist without there being a way to discern it (a particular molecule of H_2O in a body of water). Conversely, discerning identity alone does not necessarily demarcate an individual in a strict metaphysical sense (the identity of a model of car: the Subaru Forester).

17. Appiah is first analysing how identity claims work before turning a critical eye to social identities. For this reason, he begins with relatively uncontroversial examples to get at the structure. Moreover, it clearly makes good sense to pay more attention to gener-

ics that include the possibility of risk or danger. It is not that it is wrong to consider the possibility of a tiger eating you, but the question is how to include this possibility without it becoming a universal necessity.

18. Appiah cites Susan Gelman's 'Psychological Essentialism in Children' (2004).

19. See Smith 2012: 82.

20. Though I do not pursue this, it is worth noting that this conception has resonances with North American Indigenous philosophy as represented by the contemporary Indigenous philosopher Thomas Norton-Smith. See especially his discussion of constructivism in *The Dance of Person and Place* (2010: 17–19).

21. Whitehead often makes this point at varying levels of technicality. Auxier and Herstein remind us that in *The Principle of Relativity* and *Universal Algebra*, Whitehead denies that there is such a thing as 'equality simpliciter; rather, one always and only has equality with respect to some contextualizing characteristic' (2017: 95).

22. I borrow this point from de Freitas: 'Traditional philosophy of mathematics has concerned itself with the existence of mathematical objects, not the occurrence of mathematical events' (2013: 581).

23. As de Freitas writes, '… nomadic mathematics dwells in the mathematical figure as an event rather than an essence or representation. *The square is nothing but the process of quadrature*' (2013: 583).

24. This thread runs throughout Deleuze's work and is the topic of his final published essay 'Immanence: A Life'. See LS: 102–8/124–32, ATP: 261–3/319–22, PI: 27–31/4–7.

25. Richard Lewontin's work in evolutionary biology is a relevant example, though the ascription of 'novelty' in such a context remains controversial. Nevertheless, as Lewontin argues, 'organisms do not merely receive a given environment but actively seek alternatives or change what they find' (Rose et al. 2000: 313). The environment is not static, because these changes subsequently return to influence future organisms: 'Straw is not part of the bird's environment unless it actively seeks it out so as to construct its nest; in doing so, it changes the environment, and indeed the environment of other organisms as well' (Rose et al. 2000: 313). See also Lewontin 2000: 41–68.

26. Traditionally, this transcendental level refers to Platonic forms or the realm of Being as opposed to Becoming.

27. At the intersection of metaphysics and theology, this manifests most notoriously in the 'Great Chain of Being' (Lovejoy 1936). DeLanda by contrast argues that an implication of a Deleuzean metaphysics is a 'flat ontology' where 'individuals differ in spatio-temporal scale but not ontological status' (2002: 51). Deleuze does not use this terminology, though a case might be made that it follows his commitment to 'univocity'.

28. These terms are synonymous for Whitehead: '"Actual entities" – also termed "actual occasions" – are the final real things of which the world is made up' (PR: 18).

29. Harman argues that while both Deleuze and Whitehead are 'process' philosophers, only Deleuze is a 'philosopher of becoming', since Whitehead affirms the fundamental reality of actual occasions as 'fully determinate individuals' (2014a: 234). By contrast, Harman reads Deleuze and other 'philosophers of becoming' (he includes James, Bergson, Stengers and Simondon) as claiming that 'individuals per se are derivative of a more primordial dynamism' (2014a: 232). Harman is correct to see that Whitehead keeps the principle of identity in the actual occasion and Deleuze does not, but he makes too much of this difference. In particular, his manner of reading Whitehead in the tradition of occasionalism is enabled by a reliance on actual occasions as discretely serial entities that presumes a stable underlying temporal metric. Harman writes: 'Whitehead's philosophy is one in which entities are so utterly determinate that they *can last only for an instant* before perishing and being replaced' (2014a: 240, emphasis added). While Whitehead does discuss the determinacy of actual occasions as correlative with perishing, this does not imply a single temporal metric, which Whitehead explicitly denies. There is no 'unique seriality' (PR: 35); instead, durations are relative

and variable (PR: 65–6). Furthermore, given that Whitehead characterises 'eternal objects' as necessary forms for determining individuals and in terms of a potential continuity, the divergence with Deleuze is less stark than Harman makes it out to be. However, he is correct to emphasise a difference between 'becoming' and 'relations'. For a discussion of the stakes of this debate, see Bennett 2015.

30. 'Fact' is not an epistemic term for Whitehead but is instead a generic placeholder that indexes concrete and immediate reality. The stubborn fact of a situation is its reality and quality of concrete resistance or restraint. Such a stubborn fact is what it is. This is not to say that a propositional representation of a fact is equivalent to the total condition of restraint in question.

31. On the 'classical' interpretation (LeClerc, Kline are two examples) 'only concrescencing entites are actual in the full proper sense' (cited in Henning 2005: 44). When an entity has 'perished', it is dead. Jones's 'ecstatic interpretation' challenges the sharpness of the distinction between the actual occasion as concrescencing subject and as objective superject. Instead, she emphasises a continuity between subject-superject, such that the subject continues to be intensively active in future prehensions: 'the objective functioning of one thing in another . . . never completely loses the subjective, agentive quality of feeling that first brought it into being' (1998: 3). (This is the reason it is an *ecstasis*, in the sense of standing out beyond one's self.) Ultimately, the stakes of the difference hang on the underlying conception of time. The classical interpretation presumes a homogenous linear metric, whereas Jones argues that a strict delineation of past, present, and future 'operates at a separate, more abstract analytical level than the actual process of temporalization' (1998: xii). My argument clearly follows this 'ecstatic' interpretation. Henning 2005 is the best summary of the main differences.

32. The eight categories of existence are: (1) Actual Entities; (2) Prehensions or Concrete facts of relatedness; (3) Nexūs or Public Matters of Fact; (4) Subjective Form or Private Matters of Fact; (5) Eternal Objects or Forms of Definiteness or Pure Potentials for the Specific Determination of Fact; (6) Propositions or Matters of Fact in Potential Determination; (7) Multiplicities or Pure Disjunctions of Diverse Entities; (8) Contrasts or Patterned Entities. If (1) and (5) 'stand out', the other six 'have a certain intermediate character' since they could not exist without actual entities and eternal objects (PR: 22).

33. The distinction with an object-oriented inclined reading is a matter of how to emphasise this sentence. The *immortality* of this occasion is, in a certain sense, withdrawn from access. And yet, in the actualisation of further occasions, it is still vitally implicated. Note also that this is the hinge between the traditional interpretation and the ecstatic.

34. This is less of a problem for readers more interested in the rationalist procedures of Whitehead's thought, of whom Charles Hartshorne is exemplary. The most famous critique of the 'eternal objects' is John Dewey's in his 1936 APA talk in which he criticises the extent to which 'eternal objects' and 'God', which he aligns with Whitehead's 'mathematical method' (read rationalism), are in tension with his empiricism. More recently, George Allan has argued that Whitehead does not need 'God' for his system to function. In both cases (Dewey's and Allan's) the desire is for a 'naturalised' Whitehead. See Allan 2008, 2010, and Dewey 1937.

35. Whitehead adds additional nuance later: 'there are four main types of entities in the universe, of which two are the primary types and two are the hybrid types. The primary types are actual entities and pure potentials (eternal objects); the hybrid types are feelings and propositions (theories). Feelings are the real components of actual entities' (PR: 189–90). Given the non-phenomenological nature of actual entities, these hybrid types, which are hybrids insofar as they mediate between the two primary types and partake of both, are crucial for thinking existential implications of this metaphysics.

36. Whitehead's willingness to operate with these foundational categories follows his claim that 'the train of thought in these lectures is Platonic' (PR: 39), a claim that must be

balanced with his insistence that he does not mean 'the systematic scheme of thought which scholars have doubtfully extracted from [Plato's] writing' but rather 'the wealth of general ideas scattered through them' (PR: 39).

37. This statement, further evidence of the relational nature of the actual occasion, is also supportive of a reading that stresses that actual occasions are primarily metaphysical concepts, rather than equivalents of phenomenologically experienced events. To say that the actual occasion cannot be described through universals completely is to repeat the insufficiency of language for characterising experience. But this second point is more tenuous. While the constructive nature of the actual occasion, as a conceptual tool, cannot be ignored, it is also the case that there has to be some connection between this tool and what it is designed to elicit or explain, which *is* experiential.

38. '. . . the notion of intensity in Whitehead's metaphysics . . . stands at the crossroad, or indeed *is* somehow the crossroad between subjectivity and objectivity . . . the achievement of a fully determinate intensity of feeling is the purpose and nature of subjectivity . . . and this intensity is the offering of any entity as objective matter for prehension by subsequent concrescences' (Jones 1998: 85).

39. This is why Whitehead says that 'each atom is a system of all things' (PR: 36), or 'all actual entities in the actual world, relatively to a given actual entity as "subject," are necessarily "felt" by that subject, though in general vaguely' (PR: 41).

40. The source of this terminology is Husserl, but Deleuze and Guattari apply it in non-Husserlian ways. Husserl refers to the 'vague yet rigorous' or the 'essentially and not accidentally inexact' of proto-geometric forms (1931: 208). For example, the various morphological transformations which can be performed on a circle and as such are part of its essence. Deleuze and Guattari apply this conception to the concept of intensive (rather than extensive) multiplicities that are 'anexact yet rigorous . . . [and] cannot divide without changing in nature each time' (ATP: 483).

41. Whitehead defines a multiplicity as 'consisting of many entities . . . [whose] unity is constituted by the fact that all of its constituent entities severally satisfy at least one condition which no other entity satisfies' (PR: 24). Crucially, Whitehead insists that 'the only statements to be made about a multiplicity express how its individual members enter into the process of the actual world . . . *it can be treated as a unity for this purpose and this purpose only*' (PR: 29, emphasis added). The eternal objects are thus thinkable as objects only through their manner of informing actual occasions. But this does not mean that they are discrete unified objects in themselves. This use of multiplicity is another way that Whitehead's 'eternal objects' and Deleuze's 'Virtual' are closer than might first appear. As is well known, Deleuze uses the notion of a Riemannian multiplicity to characterise Virtual ideas. The resonance is in terms of the way that these multiplicities are not defined as unities in themselves, but rather offer variable conditions of determination. Dan Smith observes that 'the elements of the multiplicity are merely "determinable"; their nature is not determined in advance' (2012: 303).

42. Whitehead describes the ontological principle in different ways. Of note is his claim that it is 'the principle that everything is positively somewhere in actuality and in potency everywhere . . . thus the search for a reason is always the search for an actual fact which is the vehicle of that reason' (PR: 40). This principle constitutes 'the first step in the description of the universe as a solidarity' (PR: 40). What Whitehead is denying is that there can be a reason not manifest in some actual occasion. But the reason for this is that actual occasions are relational – which is to say that any realisation in one occasion effects, in some manner or degree, all other occasions. There are no purely private facts.

43. Decision can be thought of as selection from potentiality into actuality provided that we do not read this as necessarily an act of consciousness. The reason for this decision does not 'imply conscious judgment, though in some "decisions" consciousness will

be a factor' (PR: 43). An additional factor, discussed in Chapter 3, is what Whitehead calls 'subjective' aim. See PR: 209–15. For a helpful secondary discussion, see Krauss 1998: 75–8.

44. '[For Whitehead] the goal of creative process is the achievement of a drop of subjectivity; an event whose structure is internally self-conditioned as well as externally other-determined' (Kraus 1998: 8).

45. This divergence can be complicated in considering the phases of the actual occasion. Might not the initial multiplicity of data and the 'conformal' phase of the actual occasion be construed as a pre-individual field out of which the individual occasion emerges?

46. It is important to understand that Deleuze's use of 'Idea' is independent of a cogito or transcendental ego.

47. Infinitives are not essentially disconnected, rather every actualisation of an infinitive causes variations to ripple through other virtual infinitives: 'its determination as an infinitive is through its changing relations to other infinitives rather than as an independent entity' (Williams 2008: 128). This heuristic is particularly relevant to Deleuze's problem in *Logic of Sense*, which is focused on individuations of *sense*.

48. Though DeLanda relies on physical examples, Deleuze suggests that problematic fields and their correlative individuations are not only physical but may also be social and psychic (DR: 182–91). DeLanda's privileging of the material has given rise to an important debate with James Williams. In Williams's 2006 'Science and Dialectics in the Philosophies of Deleuze, Bachelard, and DeLanda', he argues that DeLanda's presentation of Deleuze's metaphysics as exemplified across a range of experimental and theoretical sciences 'does not allow for the ontological openness and for the metaphysical sources of Deleuze's work' (Williams 2006: 98). In not maintaining a distinction between science and metaphysics, DeLanda fails to account for the way in which a metaphysics can remain provocative even after a scientific model that it appears aligned with has been disproven. DeLanda's response to Williams's article makes clear that he does not share a constructivist image of metaphysics. He immediately objects that Williams has not revealed his own ontological commitment and wants to know whether or not Williams is an 'idealist' or a 'realist' (DeLanda and Gaffney 2010: 329). The perspective developed in this text rejects the way that DeLanda begins from stable ontological positions without considering the way these positions are produced. As I endeavour to show at length, the proper implication for a process metaphysics, developed in different ways by both Deleuze and Whitehead, is to reject this opposition (realism/idealism) as already poorly formed and begging the question. For another interesting discussion of the relation between Deleuze's metaphysics and science, see Bell 2014. For the best recent investigation of the relation between Whitehead's metaphysics and contemporary physics, see McHenry 2015.

49. *L'individu et sa genèse physico-biologique* was first published by Presses Universitaires de France in 1964, a partial publication of his doctorat d'État: *L'individuation à la lumière des notions de forme et d'information*, defended in 1958. Deleuze publishes a very positive review of this text in *Revue Philosophique de la France et de l'étranger* in 1966, now included in the *Desert Islands* collection (DI). The second part of the doctorat was not published until 1989. This complete version, under the title *L'individuation psychique et collective* is the source for the above quotations. All translations are mine.

50. In ATP, Deleuze returns to this transformation from a static 'form-matter' opposition to a 'force-material' relation inspired by Simondon. Where 'form-matter' hylomorphism 'assumes a fixed form and a matter deemed homogenous', the relation between material-force is active, 'an energetic materiality in movement' that includes 'variable intensive affects' (ATP: 408). Matter is not conceived as static, but rather as already in movement and containing 'singularities or haecceities . . . that are topological rather than geometrical' (ATP: 408): temperature, pressure, porosity, and so on as forces modulating active material.

51. Simondon's concern is driven by an interest in understanding the emergence of new kinds of technical objects without assimilating these to prior genera. Simondon is also motivated by a worry that the emerging 'cybernetics' model is insufficient.
52. Deleuze affirms in principle an infinite regress – it is difference all the way down 'ad infinitum', only qualified with the condition that this 'intensive quantum is the structure (not yet the synthesis) of heterogeneity' (DI: 87).
53. Combes gives the example of a plant: 'A plant . . . establishes communication between a cosmic order (that to which the energy of light belongs) and an inframolecular order (that of mineral salts, oxygen, etc.)' (2013: 4).
54. Versions of this debate appear throughout the secondary literature. For 'Continental' inclined readers, this is Shaviro and Harman's debate over what Harman calls 'smallism' (Harman 2014b: 41–5). In Whitehead scholarship, this issue is inherent in the split between the 'ecstatic' and the 'traditional' readings sketched above.

Chapter 3

Feeling as Creation: Affect and Tertiary Qualities

Perspectival realism and primary-secondary-tertiary qualities

The distinction between primary and secondary qualities begins in earnest during the early modern period. John Locke develops the most sustained account, introducing an additional wrinkle with tertiary qualities. Locke's analysis initiates Whitehead's reworking of this distinction, a reworking so radical that it is in effect a redefinition.

Definitions of the primary-secondary distinction are built around opposition between perceiving subject and object of perception. Primary qualities (length, width, weight, and so on) are not relative to perceiver. Secondary qualities (scent, taste, colour or sound) are. Tertiary qualities supervene on the primary or secondary, with economic values or aesthetic qualities given as examples. (As we will see, Locke's original presentation does not exactly follow this.) Such definitions incorporate constitutive assumptions of the 'bifurcation of nature'. The bifurcation as operation produces the distinction, but once posited, the distinction structures attention to reinforce the operation.

Such distinctions do capture pervasive experience of phenomenal variability. The well-known eye-of-the-beholder's variability of taste cannot be denied. But when the pervasiveness of this experience is explained through the bifurcation, we move from descriptive phenomenological distinctions to implicit (or explicit) claims about metaphysical cause. The epistemic and ontological status of the qualities then appear in order of decreasing priority: Primary qualities are the most reliable because they give us access to how the object 'really is'; secondary qualities are less reliable, since, as relational features, they are subject to variability – from striking examples such as colour-blindness to variation in perception of heat, and so on. They however still have *some* metaphysical status in referring to real outcomes of interaction between a perceiver's sense organs and some (presumably physical) qualities: while the 'sound' the vibrations of air make in striking the ear *require* the ear for their existence, the vibrations that cause these sounds do not. Finally, tertiary qualities lack almost all epistemic and metaphysical standing, due to their constitutive variability. They are relegated to the epiphenomenal, psychological or subjective.

These are, in cursory form, constitutive commitments of most 'realisms' where the existence of a mind-independent reality is reliably demonstrated by primary qualities.[1] Like the bifurcation, a mark of their entrenchment is the extent to which they appear as more or less common sense, albeit in a 'pre-critical' or 'naïve' sense.[2] Nevertheless, for all this quasi-intuitive plausibility, it is no secret that, since Kant, if not Berkeley, philosophical thought has not accepted the obviousness of the metaphysical cause of such distinctions. Kant famously denies direct access to the *Ding an sich* and thus, in some sense, even primary qualities remain mind-dependent. Berkeley also challenges the ease with which primary qualities are ascribed to a mind-independent matter where secondary qualities remain relative. The issue, for Berkeley, is the extent to which the separation of these two kinds of ideas *is an operation of abstraction*:

> but I desire any one to reflect and try whether he can, by any abstraction of thought, conceive the extension and motion of a body without other sensible qualities . . . the tenet of extended movable substances existing without the mind depends on the strange doctrine of abstract [general] ideas. (2003 [1710]: 35)

Complicating, challenging or affirming the primary-secondary distinction is at the heart of Western philosophy since the early moderns, from neo-Cartesian materialisms to transcendental or subjective idealisms (including contemporary phenomenological versions).

This is why Quentin Meillassoux identifies 'correlationism' as 'the central notion of modern philosophy since Kant' (Meillassoux 2008: 5). Correlationism names 'the idea according to which we only ever have access to the correlation between thinking and being, and never to either term considered apart from the other' (2008: 5). Inherent to correlationism is a critique of dualisms that Meillassoux dubs the 'correlationist two-step'. Not only is access to being-in-itself not possible without thought, but subject and object cannot be extricated from one another. The 'correlationist two-step' consists in the purported 'primacy of the relation over the related terms; a belief in *the constitutive power* of reciprocal relation' (2008: 5, emphasis added). Meillassoux's 'correlationism' divides philosophy and the empirical sciences, since claims about the world as it is, independent of thought, are considered naïve or reductive by the 'correlationist'.

While Meillassoux's way of setting up this problem is not without important implications, the extent to which he accepts inherited definitions of its operative terms is striking. Rather than question how the primary and secondary distinction is produced, Meillassoux declares that 'It is time [the theory of primary and secondary qualities] was rehabilitated' (2008: 1). His interest is in a realism that can make sense of what he calls 'ancestrality': empirical claims about the dating of events prior to origin of any life on Earth, but especially human life.[3] Such claims in astrophysics, geology and astrobiology challenge the 'correlationist two-step' by making claims about reality prior to the possibility of its given-ness to (human) thought: 'from the perspective of the correlationist . . . ancestral statements [cannot be] . . . interpreted *literally*' (2008: 13).

Meillassoux's real interest is not primarily epistemic. Rather he is concerned that the divide between philosophy and empirical claims about the ancestral allows for the return of fundamentalist fideisms against which thought is powerless to critique, because the 'correlationist' has destabilised rational appeal to absolute knowledge claims. He thus observes '*by forbidding reason any claim to the absolute, the end of metaphysics has taken the form of an exacerbated return of the religious . . . [such that] . . . the contemporary end of metaphysics is an end which, being skeptical, could only be a religious end of metaphysics*' (2008: 45–6, emphasis in original). The 'correlationist two-step' levels all claims into varying degrees of belief. Science becomes another variety and 'faith is pitched against faith, since what determines our fundamental choices cannot be rationally proved' (2008: 46).

Meillassoux identifies a troubling dynamic of contemporary geopolitics, but his response is less 'speculative' than it appears. Rather than challenge the operation of a bifurcation that, sometimes in the name of 'science', denies the relevance of much of experience, Meillassoux ratifies it uncritically. Primary qualities thus refer to 'all those aspects of the object that can be formulated in mathematical terms' (2008: 3) with no thought of the extent to which mathematical formulations can be expressed variably, as differential relations, as topological invariants, and so on. Construing primary qualities as the ultimate metaphysical reality begs the question in assuming that what exists is best conceptualised as stable discrete objects:

> we acknowledge that the sensible only exists as a subject's relation to the world; but on the other hand, we maintain that the mathematizable prop-

erties of the object are exempt from the constraint of such a relation, and that they are effectively in the object in the way in which I conceive them. (Meillassoux 2008: 3)

The virtue of Meillassoux's analysis is also its vice. In relying on a sharply wielded primacy of exclusive disjunction, Meillassoux assumes the universal determinacy of oppositional terms: the rational and the irrational, thought and being, and so on. There is no middle ground. This makes for strong argumentation, but bad metaphysics.

Without loosening the universal determinacy of such oppositions, attempts to challenge Meillassoux's framing are easily accused of playing the two-step. Given the basic relational structure of Whitehead's and Deleuze's metaphysics, it would not be difficult to read them as correlationists. This is impeccable argumentative strategy, but it remains entirely external to the real speculative stakes of the constructions under development. Entering into these respective constructions requires relinquishing the certainty that such broad categories (rational and irrational, thought and being) refer to determinate and fixed domains. Does this mean that we must succumb to the 'irrational' fideisms that Meillassoux is worried about? This worry is why Meillassoux wants to renew the cogency of epistemic access to 'absolute truth' and reinstall the sciences as indisputable authority. Without this,

> the struggle against what the Enlightenment called 'fanaticism' has been converted into a project of moralization: the condemnation of fanaticism is carried out solely in the name of its practical (ethico-political) consequences, never in the name of the ultimate falsity of its contents. (2008: 47)

That is, according to Meillassoux, we need to be able to say, definitively, that some content is false and some is true.[4]

But there is another way to approach the problem of fanaticisms and fundamentalisms. Indeed, in one sense Meillassoux's response is to double down on the logic inherent in all fundamentalisms, that of the one true perspective. By contrast, the implications of both Whitehead's and Deleuze's metaphysics destabilise *any* fundamentalist orientation. Meillassoux reads this as a licensing of all views as equally worthy – but this presumes that perspectivism is simply subjective. It presumes the subject invents their perspective willy-nilly. But for both Whitehead and Deleuze this is simply not the case. Perspectives are not the invention of a subject, rather, they are the forms of relational activity through which reality unfolds. Indeed, because of this structure, no single perspective can detach from this unfolding process and declare a final authority. Rather than opening up a maelstrom of fundamentalists, this shows why fundamentalisms, including scientific fundamentalisms, are reductive and partial.[5]

This is also not a giving up of the absolute per se, but rather a reconceptualising of its function. This is especially the case for Whitehead, who does not hesitate to name a conceptual operator 'God'. And yet, if it is still appropriate to characterise this 'God' as absolute, it can never be in an epistemic sense. When Whitehead conceives God 'under [an] image' that manifests a 'tender care that nothing be lost', the loss in question is not about knowledge, but rather the inheritance of feeling that has passed into 'objective immortality' (PR: 346). 'God' functions as reservoir of what has happened because these events and occasions were both not predetermined (this is Leibniz's solution) but also remain real as conditions for a future that remains open. God does not function to predict the future, but to conserve the past in a radically inclusive manner.[6] Rather than epistemic or moral, the 'tender care that nothing be lost' is a profound commitment to immanence – no matter how banal, no event or occasion is completely lost or disappeared. This entails an infinity beyond the possibility of human knowledge, but this does not make it irrational.

Meillassoux's use of 'absolute', by contrast, is always in an epistemic sense: 'we must take up once more the injunction *to know the absolute*' (2008: 28, emphasis added). This is a far cry from Whitehead's fallibilism where 'in its turn every philosophy will suffer a deposition' (PR: 7). The motivation of this fallibilism, and the hinge of the difference, is a metaphysical intuition about the nature of the real (call it the Absolutely real). For Meillassoux, however it may turn out that this real 'is', its existence is independent of human activity or thought. To say otherwise is to invoke the 'correlation'. But a process metaphysics rejects the framing by affirming both that the activity of theorising is part of the processes of reality and that this theorising can be more or less effective in how it understands these processes because they are not co-extensive. Theorising is part of the unfolding, but reality exceeds the human perspective on it. And it is because of (a) this excess and (b) the ongoing and continuing nature of the unfolding that any static claim of 'absolute' knowledge is to be viewed with suspicion. But this does not mean that 'ancestral' claims cannot function in a relatively strong epistemic sense. Nor does it mean that empirically verifiable predictions do not gain more credence than a single isolated and non-confirmed revelation. It merely keeps open the possibility that there is more to know, that we do not know it all.

Perspectivism is not primarily an epistemic claim about the equal relativity of all 'opinions', but rather part of the processual functioning of the universe. That is, it is a part of the metaphysical process that involves how the universe ('God') comes to know itself and grow. Such metaphysics do not insist on an in principle incompatibility between *religious practice* (not the same as doctrine), metaphysics and science.[7] Nor do they amount to humanist correlationism or subjective idealism, though they do reorient what 'realist' attention to experience means.[8] In this respect, Whitehead's reconstruction of the primary-secondary-tertiary distinction is crucial. Through this reconstruction we can understand Whitehead's philosophy as realist without being correlationist, though it remains relational. What changes is *what* is being related and how. In this way, Whitehead offers a differently attuned realism.

Since the basic intuitions driving the primary-secondary-tertiary distinction arise from pervasive aspects of experience, Whitehead's reconstruction cannot be exclusively phenomenological. Rather, his speculative reinterpretation involves three interrelated operations: (1) phenomenological description; (2) conceptual genealogical diagnosis; (3) speculative construction. Their combined result is an inversion of the metaphysical status of the primary, secondary and tertiary. Tertiary qualities and secondary qualities, though they no longer correspond unequivocally to ordinary psychological representation, become closer to the real. Primary qualities by contrast increase in degree of abstraction because they presuppose a 'simple location' where 'each bit of matter [is] self-contained, [and] localized in a region with a passive, static network of spatial relations' (MT: 138).[9]

This inversion is implausible without close attention to the constructions enabling it. While speculative, these constructions are linked to Whitehead's prevailing radical empiricism, where '[philosophy's] generalizations must be based upon the primary elements in actual experience as starting points' (PR: 158). Whitehead begins with a Jamesian depiction of experience as primarily a field of vague and intensive *feeling*. This phenomenological description challenges commitments to experience as built up through discrete sense impressions. Instead, what is most basic in such a field is 'blind' feeling, 'felt in its relevance to a world beyond' (PR: 163). While the feeling is 'blind and the relevance is vague' (PR: 163), this is not the blindness of Kant's intuition that requires a concept of the understanding to be resolved into coherence. Rather, the feeling already possesses a vector character, even if it is vague. It is a 'feeling from a beyond which is determinate and pointing to a beyond which is to be determined' (PR: 163).[10] While ultimately Whitehead's concept of feeling exceeds the phenomenological, for the present the challenge is to any naïve claim that primary qualities are basic elements of experience. Such claims fail to notice pervasive experiential qualities always present in some degree. Because metaphysics, as such, must be explanatorily adequate at the broadest possible degree of generality, it cannot ignore that which is always present. And yet this is precisely what most theories do in generalising from a narrow or specialised selection of experience: 'philosophers have disdained the information about the universe obtained through their visceral feelings, and have concentrated on visual feelings' (PR: 121).[11] The scale of ordinarily experienced constituted objects is not phenomenologically primary: 'those elements of our experience which stand out clearly and distinctly in our consciousness are not its basic facts; they are the derivative modifications which arise in the process' (PR: 162).[12]

This phenomenological re-description alone is not enough to enact a metaphysical inversion since it can be accommodated as psychological or perceptual only. We might grant that subjective experience begins in a feeling-laden field but enlist this field into prevailing assumptions of a substance ontology with the feelings of this field predicated of subject or self. This assumption has the appearance of clarity, but Whitehead develops a conceptual genealogy to show how it is built on inconsistent commitments.[13]

Whitehead's diagnosis demonstrates his conception of philosophy as 'the self-correction by consciousness of its own initial excess of subjectivity' (PR: 15). What he means by this is counter-intuitive to the ordinary polarity of these terms. An excess of subjectivity mistakes abstractions for concrete experience. In particular, it neglects two experiential phenomena: (1) the presence of a felt past as it informs a present; (2) the unfinished character of presence as transitional and vectorial. Denying these features, positivist empiricisms are for example excessively subjective despite claiming to begin with *only* what is given.

Correcting excess of subjectivity means tracking the ideas that generate it. The basic inconsistency is a combination of the substance-predicate metaphysical form derived from Aristotle with what Whitehead calls the Cartesian 'subjectivist principle': 'The difficulties of all schools of modern philosophy lie in the fact that, having accepted the subjectivist principle, they continue to use philosophical categories derived from another point of view' (PR: 167).[14] The substance-predicate scheme conceptualises existence in terms of stable substances qualified by predicates. Whether we are talking about essential qualities of a chair (its 'chairness' as artefact for sitting) or accidental qualities (location, colour, material, and so on) both are thought as properties inhering in an underlying substance.

The 'subjectivist principle' is the idea that 'those substances which are the subjects enjoying conscious experiences provide the primary data for philosophy, namely, themselves as in the enjoyment of such experience' (PR: 159). Whitehead has in mind Descartes's retreat, through procedures of radical doubt, to the foundational certainty of the *cogito*. More technically, the subjectivist principle is the claim that 'the datum in the act of experience can be adequately analysed purely in terms of universals' (PR: 157). Features of subjectivist experience are analysed as particulars which exemplify universal qualities: redness, roundness, hardness, and so on.

Based on this principle, experience is always tied to an experiencing subject. In conjunction with the subject-predicate metaphysical scheme, this subject is approached as a primary substance. Such a primary substance, following Aristotle, 'is always a subject and never a predicate' (PR: 157). (This remains definitional in how Descartes describes *res cogito*.) We now have to explain the relation between such a primary substance, *the cogito*, and the qualitative predicates that inhere in it. These predicates, as per the subjectivist principle, are typically analysed as repeatable universals (redness, sweetness, bitterness, and so on). So we have a repeatable universal predicated of a non-repeatable substance (sensation of greyness predicated of substance mind). What is this non-repeatable substance? It is presupposed, but difficult to explain, especially since it cannot itself be experienced.

Whitehead's analysis is largely Humean to this point. However, he adds his own tracing of its conceptual roots to the inconsistent combination of the subjectivist principle and the subject-predicate metaphysical scheme. In response, Whitehead gives up the latter (subject-predicate metaphysical scheme) and reworks the former (subjectivist principle). On his 'reformed subjectivist principle' the locus of experience is not a pre-existent subject (understood as a primary substance), *but rather the occasion of experience itself*. This is a significant speculative construction, but one that, in some sense, follows Occam's razor by not multiplying entities. What is given in experience is the occasion of experience, nothing more. Whitehead adopts the language of subject to apply to the actual occasion itself. In doing so, he denies the premise that 'divides qualities and primary substances into two mutually exclusive classes' and instead admits only 'two ultimate classes: . . . actual entities . . . and forms of definiteness' (PR: 158). The former correspond to actuality and incorporate aspects of both substance *and* quality, while the latter, with some qualification, correspond to a realm of potentiality.

Whitehead's construction, motivated to avoid the inconsistencies inherent in combining these two principles, allows him to read Locke's primary-secondary-tertiary distinction in an original way. Locke distinguished between primary, secondary and tertiary qualities to understand the experiential sources of different kinds of ideas. He links these qualities to varying 'powers' of objects of sense perception. Primary and secondary qualities are as sketched above in the opening paragraphs of the chapter. But tertiary qualities are somewhat different in referring to relational powers, whether active or passive. For example, in addition to its primary (size) and secondary (heat, light) qualities, the sun has the tertiary quality or power of melting, whereas snow has the tertiary quality of melt-ability (PR: 81, II.8.24).

Whitehead believes Locke's analysis is 'hampered by inappropriate metaphysical categories' (PR: 51). Adopting his 'reformed subjectivist principle' and rejecting the necessity of the subject-predicate scheme, Whitehead approaches qualities or 'powers' not simply as demarcating epistemic differences, but as expressing different ways actual occasions relate to, and thus contribute to, one another. The locus of these relations is no longer either the object or self-standing mind. Locke's interest in analysing the source of ideas comes with the assumption that ideas function as qualities in a self-standing substance, that is, the Cartesian subjectivist principle. Adopting the reformed subjectivist principle, Whitehead uses Locke's analysis not as applying to how human ideas represent objects in the world, but rather as indicating how one subject of experience (an actual entity) relates to others:

> If he had started with the one fundamental notion of an actual entity, the complex of ideas disclosed in consciousness would have at once turned into the complex constitution of the actual entity disclosed in its own consciousness, so much as it is conscious – fitfully, partially, or not at all. (PR: 53)

From here, the significant question is understanding how different occasions inform, constrain and relate to one another. Tertiary powers define limits and possibilities of 'how each individual actual entity contributes to the datum *from which* its successors arise and *to which* they must conform' (PR: 56). Secondary and tertiary qualities are thus *more relevant* for individuating entities that perdure over time precisely because actual occasions, as subjects of themselves, are not equivalent to macro-scale conventional objects (and subjects) of experience ordinarily conceived. Indeed, Whitehead describes such enduring 'things' as results of the 'historic' route of tertiary qualities or powers – one actual occasion imposing on the next a constellation of 'powers', that is, what it can be affected by and what it can affect: 'likeness between the successive occasions of historic route procures a corresponding identity between their contributions to the datum of any subsequent actual entity; and it therefore secures a corresponding intensification in the imposition of conformity' (PR: 56). A macro-level object is actually constituted by ongoing actual occasions that hang together through relational inheritances akin to tertiary qualities.[15]

Whitehead's speculative construction goes far beyond the phenomenological. The actual occasion is not equivalent to subjective representations. If Whitehead wants to develop Locke more *consistently*, it is metaphysical consistency he is seeking. Whitehead's point is that you need not appeal to the 'something I know not what' if you adopt the actual occasion as the inclusive metaphysical category.[16] Departing from Locke's own preoccupations, Whitehead's inversion is more accurately a redefinition, since the prevailing understanding of the three qualities is question-begging within Whitehead's metaphysics.

One consequence is a denial that there must be an underlying sameness to ground the identity of an object, idea or being. Speaking strictly, 'the exterior things of successive moments are not to be identified with each other' (PR: 55). They are not, formally, the same, since each actual occasion is its own individual.[17] Furthermore, Whitehead's analysis of actual occasions incorporates tertiary qualities with a broader category of 'feeling' as primary metaphysical operator. Feeling is 'the term used for the basic generic operation of passing from the objectivity of the data to the subjectivity of the actual entity in question' (PR: 40).[18] Actual entities are constituted through and by the way they 'feel' previous and contemporary actual entities. Indeed, Whitehead understands 'the primitive form' of all experience to be 'emotional feeling' while reminding us that 'emotion in human experience, or even in animal experience, is not bare emotion [feeling]' (PR: 163). This is to say that feeling, as generic metaphysical operation, is not just a psychological term.[19] More speculatively, feeling is not a predicate of a self-standing substance (this would repeat all of the problems leading to the bifurcation). It is rather a mode of operation that produces the actual occasion, understood as unit of satisfaction and occasion of reality. This raises two kinds of questions: (1) can we say more about the processes that lead to the occasion's achievement?; and (2) given the distinction between metaphysical feeling and ordinary conceptions of emotion, how does the metaphysical account translate into existential terms?

Subjective aim and intensity

Feeling names processes by which an actual occasion becomes the actual occasion that it is. Whitehead repeatedly insists on its 'vector' character: 'Feelings are "vectors"; for they feel what is *there* and transform it into what is *here*' (PR: 87). But in the transformation from there to here, what is being transformed and how? Since the actual occasion is, in a certain sense, in relation with the entirety of the universe, how does it not simply repeat the universe? Where does its identifying particularity come from? What makes it different? In Whitehead's words, 'How can the other actual entities, each with its own formal existence, also enter objectively into the perceptive constitution of the actual entity in question?', christening this the 'problem of the solidarity of the universe' (PR: 56).[20]

As a response, Whitehead's doctrine of prehensions includes both relations of inclusion and exclusion. 'Integrations' of the actual occasion *both include and exclude* (always by degrees) in order to 'terminate in a definite, complex unity of feeling' (PR: 56).[21] 'Concrescence' is not simply given, it is, rather, creative in the sense of adding something new. But simultaneously, this newness is not without relation to what has come before: 'The creative action *is the universe always becoming one* in a *particular unity of self-experience*, and thereby adding to the multiplicity which is the universe as many' (PR: 57, emphasis added). If prehensions include and exclude, can we say that prehension is the more abstract name given to feeling, that feeling names positive prehensions only, while the negative prehension is in some sense a refusal, rejection or otherwise inability to feel? In order to achieve 'unity of feeling', there must be a certain limiting of material, a negative prehension, a selection or omission.

And yet, this limiting or selection, in service of a process of concrescence that will ultimately add a new unity of experience to reality, must occur within a larger and more capacious feeling. If it did not, then Whitehead would have no way of explaining change or novelty in the future, since once an actual occasion had negatively prehended, it would appear that the 'object' of that negative prehension would no longer be available for subsequent prehensions. As we have seen, this is (one of) the conceptual roles played by what Whitehead calls 'God' conceived 'under [an] image' that manifests a 'tender care that nothing be lost' (PR: 346).[22] In this regard, the primordial nature of God demarcates the limits of a potentiality that is ever-present: 'viewed as primordial, [God] is the unlimited conceptual realization of the absolute wealth of potentiality. In this aspect, he is not *before* all of creation but *with* all of creation' (PR: 343).[23]

And yet, this primordial nature is not for all that unconstrained by the occurrences of actual occasions – indeed, this is what Whitehead expresses by appeal to the 'consequent nature' of God. In preserving the consequences of actual occasions, these consequences place constraints on subsequent occasions. These are the 'historic routes' that constitute the reality of particular lives and subjects (PR: 89–90). Such routes are not *a priori* necessary nor determined by the primordial nature of God. But they do constrain subsequent occasions *to some extent*. As Debaise puts it: 'each act takes up the acts that came before and transmits to those that follow' (2017a: 68). For questions of novelty and change, everything depends on the extent of the constraint. If history is contingent from a rational standpoint in not following a predetermined order, it is also constraining for the possibilities to come. Contingency has necessary effects. Openness is affirmed but also constrained by what has occurred. The consequent nature of God is 'determined, *incomplete* . . . and fully actual' (PR: 345, emphasis added).

For this reason, the stakes of negative and positive prehensions, of inclusion and exclusion, extend beyond the individual satisfaction of the actual occasion. How are we to understand the motivation of these 'choices' in the occasion?[24] Unity of feeling is the goal, but this unity is neither given nor fully imposed in the receptive phase. And yet, while Whitehead grants that the actual occasion is in one sense *causa sui* (PR: 86), in another sense it is embedded in the context and possibilities of what is given for it:

> No actual entity can rise beyond what the actual world as a datum from its standpoint – *its* actual world – allows it to be. Each such entity arises from a primary phase of the concrescence of objectifications which are in some respects settled: the basis of its experience is 'given'. (PR: 83)

Its actual world is already *limited* because this limitation is a necessary condition of its coherence (its 'order'), and yet this order is also specific to this standpoint, that is, there is no general coherence or order that applies for all time and place. Moreover, the limitation of the actual occasion's standpoint is not a full determination. Rather, the actual occasion necessarily requires an 'originative decision' and it is in this sense that it can be said to be *causa sui* (PR: 86).

This originative decision involves *how* the occasion prehends its actual world. This prehensive 'choice' is entangled with what Whitehead calls its 'subjective aim'. In one sense, all subjective aims are the same, the aim of an occasion just is its 'satisfaction' understood as the 'attainment of something individual to the entity in question' (PR: 84). But this appears circular. Indeed, as Whitehead readily notes, satisfaction is an entirely 'generic term'. *Forms* of satisfaction (or 'ways of enjoyment' (MT: 152)) vary and display 'gradation[s] of intensity' (PR: 84). We can describe the 'aim' of the occasion as a generic term for a constitutive process in every concrescence, but we cannot determine how or what this aim is in the abstract alone, beyond noting the internal drive towards 'satisfaction'. Satisfaction is in this sense its own reason or reward as achievement of a unified complex of feeling.

The issue then becomes how this satisfaction guides concrescence while still being immanent to it. Whitehead explicitly denies that 'satisfaction' can be a 'component in the process' of its *concrescence*. Indeed, satisfaction is linked precisely to the closure of this process as the effect of achieving a unity. We cannot use the goal of this achievement to explain *how* the occasion prehends or selects. Nor can this achievement be an external model or telos. Rather than a growing towards, a concrescence is 'the building up of a determinate "satisfaction" which constitutes the completion of the actual togetherness of the discrete components' (PR: 85).[25] It is a process of feeling together into a unity, but *the quality of this unity* is not given beforehand.

In the appeal to the occasion as involving a decision in relation to the settled world, there thus remains a certain inexplicability. Whitehead returns to this in *Modes of Thought*, linking it to the very question of life and a sense in which the immediacy of the occasion's self-enjoyment simply is its aim. Thinking about processes constitutive of 'life', Whitehead writes that these occasions display 'a certain absoluteness of self-enjoyment' that is its own reason (MT: 151). The self-enjoyment does not imply a subject beyond the occasion itself, rather 'each individual act of immediate self-enjoyment [is] an occasion of experience' (MT: 152). Furthermore, Whitehead notes that the intelligibility of his view 'involves the notion of a creative activity belonging to the very essence of each occasion' (MT: 151). This absoluteness of self-enjoyment is what Raymond Ruyer describes as 'absolute or nondimensional survey' (2016: 94). Like Whitehead, Ruyer understands this concept as key 'to the problem of life' insofar as it 'allows us to grasp the difference between primary consciousness and secondary consciousness' (2016: 94). For Ruyer, this absolute survey or 'absolute auto-subjective domain' is 'essentially active and dynamic' and metaphysically primary to 'spatiotemporal structures given to it' (2016: 99). For both Whitehead and Ruyer such auto-subjective enjoyment is not to be explained teleologically through reference to a further end. Rather, all ends in some sense return to this absoluteness of self-enjoyment.[26]

This raises a further question. If there remains a certain monadological structure to the very notion of absolute self-enjoyment, this absoluteness cannot be absolute isolation. Self-enjoyment is incoherent as isolation because the generation of enjoyment is a procedure of incorporating contrasts and differences into a complex unity of feeling. The greater the range of these differences, the greater the intensity of feeling achieved as self-enjoyment in the occasion. In a particularly challenging, but important, sentence, Whitehead observes: 'intensity in the *formal constitution* of the subject-superject involves "appetition" in its *objective* functioning as superject' (PR: 82).

This shows a consistent toggling between the 'private' individual satisfaction as indicated by the 'formal constitution' of the subject-superject and its public or 'objective' manifestation. The greater the intensity of range of contrasts included, the more powerful the 'objective' conditioning for future occasions: 'the heightening of intensity arises from order such that the multiplicity of components in the nexus can enter explicit feeling as contrasts, and are not dismissed into negative prehensions as incompatibilities' (PR: 83). Contrasts create the subjective occasion. As Jude Jones observes 'to be a subject is to be a provoked instance of the agency of contrast, and that is all it is' (1998: 130–1). The appetitive force of the occasion beyond itself is an effect of its achievement of intensity.

It is as if the intensity of the private satisfaction echoes through the objective or public realm and thus is more likely to condition future occasions more pervasively. If the 'self-enjoyment' which marks the achievement of the occasion is dependent on contrast, the intensity of inherited contrasts depends on the range of their incorporation of differences. Each actual occasion takes the multiplicity of diversity of other occasions and creates a new addition to this diversity: 'Each actual entity is conceived as an act of experience arising out of data. It is a process of "feeling" the many data, so as to absorb them into the unity of one individual "satisfaction"' (PR: 40).

This process of feeling the data is a 'qualification of creativity' (PR: 85). Creativity, in this sense, is the ultimate characterisation of reality, indeed Whitehead calls it 'The Category of the Ultimate' (PR: 21). The 'ultimate' of this category does not mean it is the highest aspiration, but rather that which must be presupposed by every other category: 'The Category of the Ultimate expresses the general principle presupposed in the three more special categories [existence, explanation, obligations]' (PR: 21). Whitehead unfolds this category according to three notions: 'creativity', 'many' and 'one'. One stands for 'the general idea underlying alike the indefinite article "a or an" and the definite article "the," and the demonstratives "this or that," and the relatives "which or what or how"' (PR: 21). 'Many' and 'one' presuppose each other. 'Creativity', as 'universal of universals' is the 'principle by which the many, *which are the universe disjunctively*, become the actual occasion, *which is the universe conjunctively*' (PR: 21, emphasis added). This movement, between many and one, between disjunction and conjunction, is necessary to explain why anything in the universe happens – that is, why it is not an inert static bloc.

But if feeling the data conjunctively qualifies 'creativity', what does this qualification add to the world? The satisfaction itself is over, closed and private. And yet it is also objective and immortal. Here is the rub: what passes over, what echoes on, is, for Whitehead, a 'tone of feeling': 'the tone of feeling embodied in this satisfaction passes into the world beyond, by reason of these objectifications' (PR: 85). Feeling appears as the means by which the actual occasion unifies what is *there* into a satisfaction of unity *here*, as well as the tone generated by this achievement of unification. This tone is what is taken up, subject to further modifications, in ongoing qualifications of future actual occasions.

We now return to the question of realisms and tertiary qualities. Understanding Whitehead's philosophy as a perspectival realism hangs on understanding that the *public/private polarity is inherent in all occasions of existence*, not just those occasions consisting of what we ordinarily think of as sentient consciousness. This is also to say that 'mental activity is one of the modes of feeling belonging to all actual entities *in some degree*' (PR: 56, emphasis added). It is sentences like these which enable characterisations of Whitehead's philosophy as 'panpsychism'.[27] From Meillassoux's perspective, this disqualifies any claim to a realist philosophy and exhibits key features of what he calls 'strong correlationism'. A strong correlationist metaphysics is

> characterized by the fact that it hypostatizes some mental, sentient, or vital term: representation in the Leibnizian monad; Schelling's nature, or the objective subject-object, Hegelian Mind; Schopenhauer's Will; the Will (or Wills) to Power in Nietzsche, perception loaded with memory in Bergson, Deleuze's Life. (Meillassoux 2008: 37)

In this case, everything depends on thinking by degrees. As Whitehead stresses, the model for what he calls 'mental activity' or 'conceptual feeling' *is not human consciousness or sentience*. Rather, such sentience represents a particularly specialised and exceptional version of this conceptual feeling ('conceptual feelings do not necessarily involve consciousness' (PR: 165)). It is a question of willingness to speculatively create a concept not captured by preconceived ideas of the split (public/private, mental/physical) in question. Many if not most actual occasions exhibit minimal 'conceptual feeling', but minimal is not necessarily zero, as Hartshorne puts it.[28]

Is Whitehead hypostatising a sentient term and ascribing 'consciousness' to everything? This question can just as easily be turned around. Choose your poison. Meillassoux could be accused of hypostatising the concept *matter*. Though ordinary modes of thought ('common sense') take matter as self-evident, they are structured by what Whitehead terms the fallacy of simple location in presuming that

> it is adequate to state that [a bit of matter] is where it is, in a definite finite region of space, and throughout a definite finite duration of time, apart from any essential reference of the relations of that bit of matter to other regions of space and to other durations of time. (SMW: 58)

In a consistent thread of Whitehead's work he insists that such a conception of matter is no longer consistent with physics. Because it is based on 'the notion of continuous stuff with permanent attributes . . . retaining its self-identity through any stretch of time' (PR: 78), this (metaphysical) conception of matter makes it difficult to think the activity of quantum physics. It is a conceptual relic of the past which continues to operate in the present and in ordinary materialisms.

Given Meillassoux's emphasis on the need for philosophy to legitimise the notion of science as providing absolute (rather than correlative) knowledge, it is odd that he appears content to implicitly operate with a definition of matter as inert and static. But, in a twist of ironic fate, he is in good company here, since it is Kant, the original 'correlationist', who defines matter in this way: 'The concept of [lively matter] involves a contradiction, since the essential character of matter is lifelessness, inertia' (1987 [1790]: §73, 276/395).

Choose your presumption then: death or life, stasis or creativity?[29] If we insist on disputing with such broad abstractions, the question remains a vicious circle. Part of the effect of Whitehead's and Deleuze's speculative philosophies is to destabilise the assumption that these terms have a stable meaning that is metaphysically *neutral*. Of course, it is not difficult to cite passages in both to accuse of panpsychism or vitalism, if we presume those terms mark stable positions. Consider this well-known passage from *Difference and Repetition*:

> What we call wheat is a contraction of the earth and humidity, and this contraction is both a contemplation and the auto-satisfaction of that contemplation. By its existence alone, the lily of the field sings the glory of the heavens, the goddesses and gods – in other words, the elements that it contemplates in contracting. What organism is not made of elements and cases of repetition, of contemplated and contracted water, nitrogen, carbon, chloride, and sulphates, *thereby intertwining all the habits of which it is composed.* (DR: 75, emphasis added)

Displaying Deleuze's characteristic poetic rhetoric (heavens, goddesses and gods), this does not amount to an unequivocal statement of 'vitalism' unless we take it as a self-standing statement of ontology. More important than 'vitalism' is how the organism is made up of actions of 'contemplation' and 'contraction' – actions oriented by an 'auto-satisfaction' that echoes Whitehead. It is not so much an ascription of something called life to the elements, but a reconceptualisation of life as habits of repeating relational activities. This displays the *ontological function of habits or contemplations* in the case of an organism, but contemplation and habit are broader metaphysical categories ('habit here manifests in its full generality' (DR: 74)). They are not necessarily activities *of* the organism, but the activities by which an organism is constituted or composed. We can get smaller and smaller in thinking these activities of habitual contraction, of the 'thousands of passive syntheses of which "we" are organically composed' (DR: 74). At some point, the border between the living and the non-living is blurred and no longer primary metaphysically. What is primary is the activity of habit, contemplation and contraction. None of these requires a (human) mind or an organism. This is a different way of approaching matter, not an incorporation of matter into mind or of material into life.

Undoubtedly, this philosophical merry-go-round will keep spinning, especially if we continue to operate through concepts presumed determinate (mind, matter, spirit, rational, irrational, life, death) and leveraged in an oppositional dialectic. The question here is how *thinking by degrees* – coupled with a conceptual turn to making actions, processes and events primary – changes conceptions of experience. The status of tertiary qualities is one way to emphasise these effects.

I claimed above that tertiary qualities can be thought as 'more real' than 'primary qualities' because they are closer to the operative feelings by which actual occasions 'feel' their context and achieve their satisfaction. But after closer consideration of the metaphysical operation of feeling, this claim must be qualified. Indeed, failure to qualify this claim opens it to a charge of subjectivist relativism. Contemporary usage of tertiary qualities practically guarantees this charge since they are defined as relative to human pleasure. Locke's original concept does not, however, since tertiary qualities need not have any human relation. They mark 'powers' of one 'object' in relation to another. Whitehead shifts the locus of Locke's analysis from self-standing objects and minds and replaces them with the 'actual occasion'. What exists are actual occasions. Tertiary qualities are descriptions of how one occasion objectifies the diversity of its context *and* the effects that this has on contexts for future occasions.

How does this translate into experience of tertiary qualities? How we answer this question depends on willingness to understand the reality of experience's excess to the language of its representation. It depends on a capacious scope to the term experience, in which we may be living realities we are not conscious of, and yet which have real effects on our lives. For example, we might say the sun-snow relation is characterised by tertiary qualities of meltability/melt-provoking (heat). This is clearly in one sense true, but it is also a significant abstraction. But these are abstractions by which we represent a *real potential relation* in the actual occasions constituting snow and sun. Do we experience this real potential relation directly? This question is why Elizabeth Grosz repeatedly emphasises the asubjective, non-phenomenological nature of sensations, affects and intensities:

> What differentiates them from experience, or from any phenomenological framework, is the fact that they link the lived or phenomenological body with cosmological forces, forces of the outside, that the body itself can never experience directly. Affects and intensities attest to the body's immersion and participation in nature, chaos, materiality. (2008: 3)[30]

Deleuze's distinction between difference and empirical or qualitative diversity is useful here, since it complicates reading Whitehead's 'feeling', and by extension, tertiary qualities, as directly equivalent to ordinary modes of representation. As Deleuze puts it, 'Diversity is given, but difference is that by which the given is given, that by which the given is given as diverse' (DR: 222). The diversity of phenomenal qualities, what we perceive, both secondary and tertiary qualities, are manifestations of metaphysical difference: 'Everything which happens and everything which appears is correlated with orders of difference: differences of level, temperature, pressure, tension, potential, *difference of intensity*' (DR: 222, emphasis in original).

For this reason, Deleuze declares that 'intensity is the form of difference in so far as this is the reason of the sensible' (DR: 222). Intensity cannot be rendered in the direct language of qualities because qualities already have cancelled out or explained the differences in intensity that produce them. In a sense, any stable phenomenal quality is a sign that is composed of series of differences. The eye sees red, but this red 'signs' a series of heterogeneous relations, light, pressure, neural charges, and so on. The quality of the red varies as intensive degrees change, each of which has differential effects on other variables. Series reverberate and extend, 'each intensity is already a coupling . . . we call this state of infinitely doubled difference which resonates to infinity disparity' (DR: 222).[31] Thus, any static rendering of a tertiary quality is not sufficient: 'the reason of the sensible, the condition of that which appears, is . . . the Unequal itself, *disparateness* as it is determined and comprised in differences of intensity, in intensity as difference' (DR: 223).

Deleuze's rendering of *disparateness* is closely related to Whitehead's ultimate of ultimates, creativity. Disparateness and creativity are not experienced in themselves, though everything that is experienced is a qualification of disparateness or creativity, in varying degrees or intensities. Whitehead's feeling can be thought through the language of tertiary qualities only if we de-subjectify these qualities. Put in psychological terms: anger, fear, pleasure, sorrow, joy are human terms that explain real occasional effects, but the feelings of these occasions extend beyond the confines of the human. This is not to ascribe anger *qua* human emotion to nonhuman occasions, it is to explain anger *qua* human emotion as *representing* a real feeling of occasions.

The distinction between intensive and extensive features is also useful here. Deleuze and Whitehead both use this technical distinction to complicate empiricisms. Both intensive and extensive features are, in one sense, experienced. But only the latter are able to be represented.

> It turns out that, in experience, *intensio* (intension) is inseparable from an *extensio* (extension) which relates to the *extensum* (extensity). In these conditions, intensity itself is subordinated to the qualities which fill extensity (primary physical qualities or *qualitas* and secondary perceptible qualities or *quale*). In short, we know intensity only as already developed within an extensity, and as covered over by qualities. (DR: 223)[32]

My earlier claim can be restated with more precision. Human experience of tertiary qualities is more real if taken as a sign (not an equivalent) of the metaphysical operations of feeling constitutive of actual occasions. This means they must be interpreted and are not directly self-certifying in any normative sense. But it still follows that primary qualities are not the most concrete expressions, but rather are more abstract. In this sense, while still real, they are less intensely implicated in processes of concrescence constituting actual occasions.

Attention, attunement and becoming-imperceptible?

How does Whitehead's metaphysical 'feeling' relate to the concept of 'affect' so important to Deleuze and Guattari in *A Thousand Plateaus*? As is well known, Deleuze's and Guattari's use is derived primarily from Spinoza with affect referring to the 'ability to affect and be affected' (ATP: xvi).[33] At first, it appears that the two discussions are incompatible, since discussion of affect occurs at the level of bodies, whereas Whiteheadean feeling is prior to bodies, that is, feeling is that by which bodies do or do not cohere.

Deleuze and Guattari follow Spinoza in asking 'what can a body do?' (ATP: 256) and thus appear to presuppose the body as existent entity. But this reading presumes habits of a substance-predicate ontology and is easily complicated. For one, bodies themselves are described as consisting of both 'latitude' and 'longitude': 'Latitude is made up of intensive parts falling under a capacity and longitude of extensive parts falling under a relation' (ATP: 256–7). Latitude is more important for affect: 'the latitude of a body [is] the affects of which it is capable at a given degree of power' (ATP: 256). Affect thus refers to the shifting *capacity or power* of intensive parts of bodies in relation. The use of 'intensive part' here is confusing, since parts are more readily conceivable in relation to extension. But it appears that the reference has to do with the extent to which a body is neither exclusively extensive nor intensive. As such, 'the latitude of a body' refers to 'affects of which it is capable at a given degree of power' (ATP: 256). Intensive parts refer to degrees of power and different degrees mean different intensive parts: while the longitudinal relations of the body refer to its varying contact with other bodies, latitudinally there is the body under duress, under joy, under sorrow, and so on. Consider the intensive degree of the eye under conditions of intense labour or in the precision of a desert dawn.

Deleuze and Guattari also state that 'affects are becomings' (ATP: 256). It is not that the body *has* an affect but rather that affect conditions and shapes bodies in dynamic processes. The 'same' body becomes different under differing affective conditions, up to a limit or threshold at which point changes in intensities push the body's arrangement past a current form of recognition.[34] This relational or affective capacity is ontologically prior to categories of species and genus or metaphysical forms or essences. A body is defined by 'counting its affects', not locating its essence, and, as such, affects cut across traditional taxonomies: 'a racehorse is more different from a workhouse than a workhorse is from an ox' (ATP: 257). Indeed, on what Deleuze and Guattari call the 'plane of consistency' – which corresponds to reality prior to its consolidation and representation into macro-level entities and objects – there are 'nothing but affects and local movements, differential speeds' (ATP: 260).

Affect, like Whitehead's feeling, is thus also metaphysically prior to objects and bodies that it conditions. Indeed, Deleuze and Guattari go further in discussing *haecceity* as a mode of individuation 'very different from that of a person, subject, thing, or substance' (ATP: 261). Haecceities 'consist entirely of relations of movement and rest between molecules or particles, capacities to affect and be affected' (ATP: 261).[35] Such haecceities 'have a concrete individuation' that 'direct the metamorphosis of things and subjects' (ATP: 256). Deleuze and Guattari offer several indications of what such a mode of individuation means experientially. Think of the individuality of a 'degree of heat', an 'intensity of white', a certain gust of wind. In a sense, we can think of haecceities as an effort to develop an existential language for a pre-personal processive view of individuations in experience. This is why they say that 'the individuation of a life is not the same as the individuation of the subject that leads it' (ATP: 261). A life, in its precise haecceity, is excessive to how the subject narrates its meaning. This is why abrupt shifts and changes in understanding are possible. Another way of putting this, in hybrid language with Whitehead, a life is an abstract name for a relative consistency in the qualifying feelings of actual occasions.

Deleuze and Guattari differ from Whitehead, however, in valorising the possibility of existential experience of haecceity, even flirting with making it normative: 'For you will yield nothing to haecceities unless you realize that that is what you are, and that you are nothing but that' (ATP: 262). Such realisation alters orientation towards the events of life:

> You are longitude and latitude, a set of speeds and slownesses between unformed particles, a set of nonsubjectified affects. You have the individuality of a day, a season, a year, a life (regardless of its duration) – a climate, a wind, a fog, a swarm, a pack . . . *Or at least you can have it, you can reach it.* (ATP: 262, emphasis added)

But why would one want to reach this? Deleuze and Guattari have just taken pains to insist that the subject and a life are distinguishable as manifestations on different planes. A life occurs on the 'plane of consistency or of composition of haecceities which knows only speeds and affects', whereas the subject lives on the 'plane of forms, substances, and subjects' (ATP: 262). How do these two planes intersect?

Clearly, Deleuze and Guattari valorise the plane of consistency, but everything depends on how this valorisation proceeds. It cannot be a matter of making one plane 'real' and the other 'illusion', though they flirt with this in some of their language. It is a question rather of living your individuation process to its greatest intensity. Insofar as one does not realise the plane of consistency and its haecceities, one remains insulated from this intensity. But this cannot mean simply dismissing the plane of forms and substances, can it? Rather, how do we navigate both planes in a way that does not diminish or reduce the intensities potential in haecceity?

Haecceities as modes of individuation means construing *subjects as events*:

> the wolf, the horse, the child . . . cease to be subjects to become events, in assemblages that are inseperable from an hour, a season, an atmosphere, an air . . . The street enters into a composition with the air, and the beast and the full moon enter into composition with each other. (ATP: 262)

What is individuated by haecceities differs from ordinary demarcations of a substance ontology. *The event is the assemblage is the individual haecceity.* Separations into constituent entities are effects of analysis. Ontologically, it is the whole assemblage that is taken as the event of individuation: 'climate, wind, season, hour are not of another nature than the things, animals, or people that populate them, follow them, sleep and awaken with them' (ATP: 263). The plane of forms and substances separates, through procedures of analysis, what originally comes together. Deleuze and Guattari suggest that realising this togetherness has positive 'affects' – that is, it increases or augments the power of the body that realises it. At once, however, and paradoxically, the condition of this augmentation is a realisation that one does not possess power as an independent sovereign because any such independence requires separation and denial of the assemblage of haecceity.

A life *is* its relational events, and the quality of the affect garnered, transmitted and conducted through these events varies according to capacity to realise their haecceity: 'This animal is this place! . . . That is how we need to feel' (ATP: 263). It is not you and the music, it is: *you-listening-to-the-music*. That's the haecceity, that's the event. But already this example raises worries, as if there were just 'you' and the 'music'. What about all of the other 'yous'?

If the goal of individuation is separating and determining precise boundaries between one entity and another, then haecceity as mode of individuation must be rejected. This is ironic, since haecceity can also be thought of as the most precise and singular mode of individuation possible, but its realisation challenges the determinability of precise boundaries generalisable from any abstract perspective. *Haecceity cannot serve the interests of property or possession.* As Deleuze and Guattari say: 'A haecceity has neither beginning nor end, origin nor destination: it is always in the middle. It is not made of points, only of lines' (ATP: 263), which is to say tendencies, directions and actions in process.

Experientially, it does seem that the song comes to an end, the concert finishes, the band leaves the building. And yet its echoes continue. We eat the last morsel, but when does the dinner end? Is it when the last molecule is integrated into the eating body? Notoriously then, an event-based ontology seems unable to distinguish one event from the other, they bleed together, we are always in the middle, never arriving. Again, the question returns: why do Deleuze and Guattari say that this is 'how we need to feel' (ATP: 263)?

This question occupies the bulk of Part II. I end the chapter with a few remarks as promissory notes. Taking this seriously heightens stakes of attention. Sensibility, affective range of capacity, attention to specificity and intensity are manners of activity that shape individuated moments. Moments are synthesised through feeling forms that bring together a range of disparate data into a unique experience. While these feelings alone are not fully determinate, they are also not merely subjective in a psychological sense. Or rather, they are subjective, but the locus of the 'subject' is the occasion, not a macro-organism.

Alterations in attention have the potential to transform an individuating moment, because attention is a crucial component enacting the form of the occasion.[36] Since each actual occasion is in relation to all others in gradations of relevance, intensity and importance, *then what gets left out partially shapes the form of what gets included.* Neither is independent of the other. Responding to this dynamic emphasises the relation between the speculative and the experiential: can a shift in attitude towards this relation alter the 'decision' of actual occasions? Actual occasions are not directly equivalent to experiential events as these are consciously represented. They are rather, in a sense, the condition of possibility underlying the conscious understanding of the events of life as they occur. Actual occasions are continuously occurring below the threshold of awareness as the constitutive conditions of that awareness. But here we must inject a further caution: actual occasions are themselves conceptual constructions designed as more coherent means of thinking reality. As an abstract frame then, there is a way in which, provided we don't make it an equivalent, the actual occasion is flexible enough to also apply to experiential events. At the very least, there can be no essential gap between experiential events and actual occasions.

For this reason, what James Williams diagnoses as the two problems of an empiricist representation of experience become centrally problematic for thinking the stakes of attention. These are what he calls 'empirical oblivion' and 'latent significance' (Williams 2010). The former refers to the necessary limitation inherent in any representation of an event: 'any record of an event involves forgetting aspects of that event which may later turn out to be of great significance' (Williams 2010: 25). The latter is the possibility that precisely this forgotten aspect may re-emerge in the future as significant. For Deleuze, these problems are correlative: 'neither is complete without the other and any real event is both an event of erasure and a novel reawakening' (Williams 2010: 30). Both problems are inherent in the intersection between attention and actual occasions. In understanding attention as always implicated in erasure (or limitation) and the possibility of emerging significance, could a more capacious attention alter the selections of occasions? This conceptual question does not translate directly to an intentional operation: we cannot simply *choose* to notice what we have not noticed in the past.[37]

To what extent can attention, differently motivated, become attuned to more expansive ontological levels of occasions? Another way of putting this is to what extent can attention, differently motivated, change the subjective form of the actual occasion? The implications of this question are twofold: more modestly, a shift in subjective form is a shift in feeling that brings features or elements of actual reality into view that were previously unseen. This can inspire a greater awareness of embeddedness in relations excessive to narrow self-interests. In this sense, it may inspire what I call an ecological attunement. More radically, a shift in subjective form, in shifting the concrescence of actual occasions, changes reality. *It changes what becomes real.*

The distance between the modest and the radical implications is the distance between the phenomenological and the speculative. The bridge between these depends on exploring affect or feeling without presuming these are fully captured in the language of emotional states. Understanding feeling as not fully dependent on a subject, that is, as not a predicate of a substance, encourages an attention to dynamic patterns of energetics that can be a-subjective: the rigorous rush of wind through the boughs of the pine, the slow outward rippling of concentric circles after a single stone disrupts the surface of the pond.[38] An ecological attunement under-stands such kinetic dynamisms to be more than just pretty pictures, but expressions of interdependence out of which subjects arise. Deleuze's and Guattari's 'becoming-imperceptible' has this kind of attunement in mind, an attunement to the molecular presence of events.

Metaphysical accounts of affect and feeling also help explain how expe-rience of tertiary qualities is *both* highly variant and yet *still* responsive to an extra-subjective real excessive to the subjective perceiver. They help make sense of the intuition that one's perception of, say, the charged ambience of the forest at twilight as the light shifts and the shadows lengthen, is not merely a trick of the subjective mind, even as this verbal description inevi-tably simplifies the actual relations it expresses. Something real drives each auto-satisfaction of occasion, beyond only the primary qualities typically conflated with physical reality. This encourages us to pay more attention to the affective feel of a place, to not dismiss this as 'just in our head'. As Whitehead enigmatically puts it: 'We experience the green foliage of the spring greenly' (AI: 250). This does not mean the descriptions through which these feelings are shared are *normatively conclusive* in any traditional epistemic sense. But it does mean that they are expressive of a reality that is not purely subjective in the sense of the human subject alone. The greenly experienced foliage is an experience of something real beyond the boundaries of its human representation. When artists express the beauty of the greenly felt spring, this expression is shared with other occasions that make up that spring. The greenly felt subjective form informs other actual occasions which in turn cluster into other societies. Whitehead's structure cuts across a presumed opposition between the merely private and subjec-tive and the assumed objective or normative. Satisfactions of occasions are at once private and singular and public and objective. The quality of the feeling exceeds the feel-er.

The feeling at the top of the mountain encountered by the human interloper is not just a trick of human psychology, though access to the feeling is modulated through the prevailing habits of that psychology. But there is a feeling there – or, more accurately, there are many feelings there – that drive the ongoing coalescence of actual occasions. Deleuze writes that 'the landscape *sees*', using language to express something that is not technically 'a' landscape, as unified subject, 'seeing', but rather *the myriad occasions of feeling that the landscape is* (WIP: 169). The feelings that move through one in these events are also, potentially, part of the landscape's seeing. This is not a claim of identity, it is one of participation: 'We are not in the world, we become with the world; we become by contemplating it' (WIP: 169). Feelings change as relations shift, and indeed the 'entities' of the mountain top are consistently unfolding new implications of the past route of occasions driven vectorially through feeling. The wind, the changing weather, the arrivals and departures of macroscopic organisms are all factors in the synthesis of each specific and singular actual occasion at this nexus. A first step in becoming more attuned to such feelings, and thus becoming more attuned to a wider range of relations beyond the narrowly anthropocentric, is to acknowledge their reality. To thinkers of cultures in which this reality was never in doubt, this might seem like a flimsy conclusion. But given the degree to which the primary-secondary-tertiary structure remains entrenched in Western onto-epistemology, these arguments through the categories of Western philosophy may offer alternative routes towards related realisations.

Notes

1. Though a process metaphysics challenges the assumption that this is an exclusive disjunction, the terminology of realism and anti-realism is pervasive in contemporary philosophy. Braver 2007 is largely consistent with Meillassoux in narrating the history of Continental philosophy through the language of realism and anti-realism to put it in dialogue with analytic philosophy.
2. It is additionally striking that their apparent conceptual ease does not match their lived fervour. Disagreements are not launched over the length of the painting, but rather where it looks best.
3. Meillassoux refers to two techniques: (1) those using radioactive decay, most typically radiocarbon, and (2) techniques known as thermoluminescence dating that can determine the time elapsed since a sample was exposed to sunlight. He cites Roth and Pouty 1985. Meillassoux is primarily interested in the possibility of ancestral claims, not scrutiny of how they are technically produced.
4. Meillassoux connects the sceptical attack on absolute knowledge with a return of fundamentalism: 'Having continuously upped the ante with skepticism and criticisms of the pretensions of metaphysics, we have ended up according all legitimacy in matters of veracity to professions of faith – and this no matter how extravagant their content' (2008: 47).
5. The use of scientific fundamentalism is risky in the context of contemporary politically motivated attacks on science. I do not mean to open up a complete jettisoning of scientific consensus. I am rather referring to a rhetorical tendency to presume 'absolute' knowledge where in fact science remains, in principle, fallible and probabilistic. My reference to scientific fundamentalism therefore refers to a reductive view of science itself.
6. Whitehead develops a bipartite analysis of what he calls God's 'primordial' and 'consequent' nature (PR: 343–51). The primordial nature of God grounds the potentiality of all reality. However, the consequent nature expresses the extent to which this potentiality remains open and not predetermined. The occurrences of actuality leave traces on the conditions for future occasions that are both contingent but constraining. Whereas the primordial nature of God is aligned with pure potential, the consequent nature of God is 'determined, *incomplete* . . . and fully actual' (PR: 345).
7. Meillassoux conflates religion and theology in a caricature of the 'irrational' or 'fideist'. Moreover, the question of faith is always epistemic for him – it is not what one does with belief, but only the way that faith for him destabilises knowledge. Whitehead by contrast has a much more empirically adequate characterisation of religion as exhibiting four factors: ritual, emotion, belief and rationalisation (RM: 18).

8. Despite the intuitive appeal of Meillassoux's reference to the mathematisable features of 'an object' as foundation for realism, Whitehead reminds us that 'quantity' is not the root of mathematical reasoning, but rather 'pattern': 'beyond all questions of quantity, there lie questions of pattern, which are essential for the understanding of nature' (MT: 143). For this reason, Meillassoux's use of ancestral claims as an example of 'absolute' empirical claims are thin, since they presume a single metric order of temporal magnitude as absolute.

9. Whitehead's diagnosis of the 'fallacy of misplaced concreteness' or 'the fallacy of simple location' is a consistent thread throughout his work. See, for example, SMW: 51, 58; PR: 137; MT: 137–40.

10. Both Shaviro 2014: 77–84 and Stengers 2011: 202–3 also recognise the importance of James's 'pure experience' for this claim in Whitehead.

11. Whitehead on this point is closer to Bergson than the classical pragmatists as they are typically read, since he insists that the habits of the organism in experience are to some extent to be overcome in the interests of philosophising. See Allen 2013; and MT: 106.

12. This claim relies on Whitehead's distinction between two modes of perception: 'causal efficacy' and 'presentational immediacy'. The latter refers to sense perception as ordinarily conceived and is the sole form of perception acknowledged by ordinary empiricism. The former refers to the visceral felt sense of being constrained by conditions of physical reality. This analysis is prominent in *Symbolism* and PR (168–83). In *Adventures of Ideas* 'causal efficacy' is referred to as 'nonsensuous perception' (AI: 182–4). In *Modes of Thought*, the distinction is described as the contrast between 'sense perception' (presentational immediacy) and 'primitive' bodily experience (causal efficacy) (MT: 72–6, 112–17).

13. Though I use 'genealogical' in unorthodox fashion, it is consistent with the impetus behind typical usage. If Foucault, following Nietzsche, tracks historical and social conditions of possibility of a given episteme, Whitehead offers a metaphysical genealogy that is not the less historicised. That is, he tracks the structuring influence of metaphysical assumptions with specific historical roots in particular thinkers.

14. There is also the idea that we passively receive sense impressions rather than construct or shape them in some active fashion. Whitehead calls this the 'sensationalist principle': 'the primary activity in the act of experience is the bare subjective entertainment of the datum, devoid of any subjective form of reception' (PR: 167).

15. Whitehead calls such conventional objects and subjects 'societies'. I investigate his reasons and manner of doing so at length in Chapter 4.

16. 'If anyone will examine himself concerning his notion of pure substance in general, he will find he has no other idea of it at all, but only a supposition of he knows not what support of such qualities' (Locke, II.XXIII.2).

17. The quality of sameness is located only within what Whitehead calls 'societies' – see Chapter 4.

18. Whitehead's reading of Locke transmutes Locke's conception of 'idea' into this operation of 'feeling': 'Its [the formal entity] ideas of things are *what* other things are for it. In the phraseology of these lectures, they are its "feelings"' (PR: 51).

19. Nevertheless, while the psychological experience of emotion is not equivalent to (metaphysical) feeling it is still that which 'most closely resembles the basic elements of all experience' (PR: 163).

20. Jorge Nobo declares this solidarity to be the 'fundamental thesis of Whitehead's metaphysical philosophy' (1986: xiv) and its most fundamental problem. How do we understand this solidarity without positing an underlying stable substratum? That is, how is this solidarity both produced and fundamental?

21. Whitehead notes that the 'classical doctrine of universals and particulars, of subject and predicate, of individual substances not present in other substances, of the externality of relations, alike render this problem incapable of solution' (PR: 56).

22. When Whitehead reminds us that this 'is but an image', he highlights the extent to which his 'God' is not anthropomorphic (PR: 346). This tender care and 'infinite patience' are conceptual features, not characterisations of an agential deity.

23. In Deleuzean terms, this is a commitment to immanence rather than transcendence.

24. In his 1987 lecture on Whitehead, Deleuze adopts his own terminology for thinking about this: in what sense is the genesis of an actual occasion a 'conjunction' and how is this genesis related to the 'disjunctive diversity' that is its condition? Like Nobo, Deleuze understands what he calls 'convergence' (Whitehead's 'objectification') as prior to 'conjunction' (his term for the genesis of the actual occasion akin to 'concrescence'): 'the first stage [is] the many or the disjunctive diversity . . . the third stage, the formation of series converging towards limits . . . the fourth stage . . . the actual occasion, it is the conjunction. *The conjunction comes after the convergence*' (emphasis added, my translation). The lecture, which occurs during Deleuze's seminar in preparation for *The Fold*, is notable for the attendance of Isabelle Stengers. Deleuze announces his discussion of Whitehead as inspired by Stengers's presence: 'I need some help [with] certain problems of physics . . . Isabelle Stengers is here today, and she will not be here the other weeks, [we should] profit from her presence' (http://www.webdeleuze.com/php/texte.php?cle=140&groupe=Leibniz&langue=1 (my translation)). Unfortunately, Stengers's response to Deleuze's questions are frequently inaudible and not included in the transcription.

25. Whitehead analyses this 'building up' into phases. Understanding the status of these phases is a challenge. They cannot be temporal, linear or spatial *parts*, since this introduces a field pre-existing independently of the occasion. As Debaise notes, one source of confusion is a tendency to confuse the potential for analytic division with actual division. The actual entity or occasion, as Whitehead stresses, is a unity insofar as it is an achievement or activity of *unifying*. Nevertheless, if it can be analysed into parts, we should not understand this analysis to determine parts that exist separately. The mistake is that we substitute 'the possible for the real': 'Because a totality can be divided we conclude that it is in fact actually divided, composed of distinct and divisible parts' (2017a: 112).

26. Deleuze also adapts Ruyer's 'absolute survey' (WIP: ix, 209–11). Ruyer frequently notes Whitehead's influence in *Neo-Finalism*, and his description of 'absolute survey' as 'form-in-itself of every organism and at one with life' (2016: 98), is clearly consistent with Whitehead's 'satisfaction'. In both cases the effect is to posit an absoluteness that 'dispenses with infinite regress' (2016: 92). However, while Whitehead's actual occasion remains fundamentally speculative, Ruyer uses 'absolute survey' in the service of phenomenological analysis of perception. See Ruyer 2016: 90–103.

27. Such characterisations can be sympathetic (Shaviro 2015) or dismissive (Rorty 1979). While they have some credence, they risk discouraging the need to consider the details of how Whitehead develops his view. Because 'panpsychism' is an easily articulated view, it is also easily denounced. 'They're saying that rocks have consciousness, whoever heard of such an absurdity! These philosophers will believe anything.' This is not to say there is no place for engaging Whitehead's thought through this lens, only that I find the term (which Whitehead does not use) to not be worth the risk.

28. Hartshorne formulates what he calls the 'zero fallacy' as follows: 'with properties of which there can be varying degrees, the zero degree, or total absence, is knowable empirically only if there is a known least quantum or finite minimum, of the property'. For example: 'A zero of elephants is observable because there is a finite minimum of what can properly be called an elephant' (Hartshorne 1997: 166).

29. Part III 'Nature and Life' of Whitehead's *Modes of Thought* outlines the fundamental stakes of this choice in emphasis. In 'Nature Lifeless', Whitehead discusses what he calls the 'Hume-Newton situation' as the 'primary presupposition for all modern philosophic thought' that results in what he considers an incoherence: 'a dead nature can give no reasons' (MT: 135). In 'Nature Alive', Whitehead discusses an alternative,

speculative approach that reframes the presuppositions in a manner consistent with modern physics. See also McHenry 2015.

30. In separating 'affects' from 'affections' and 'percepts' from 'perceptions' we see the complicated dance with Kant, in which the *noumenal* becomes forceful – in which there is no transcendental unity of apperception, but in which the form of the transcendental argument still lingers (Grosz 2008).

31. Deleuze's description of intensity as an infinitely doubling differential series ('Every intensity is E-E', where E itself refers to an e – e', and e to ε – ε' etc.' (DR: 222)) closely follows his 1987 characterisation of Whitehead's discussion of the event in *Concept of Nature*. He is interested in a characterisation of the event as an infinite series with no final term, but with a definite limit. The event would therefore have a qualitative characteristic (a vibratory harmonic) which cannot be exhausted in a single determination. See: http://www.webdeleuze.com/php/texte.php?cle=140&groupe=Leibniz&langue=1

32. The Latin terms (*extensio* and *extensum*) refer to the act or process of extension and its result. Deleuze uses *extensité* or *extension intra* for the former, and *étendue* for the latter. See Paul Patton's note, DR: 329.

33. In Part III of the *Ethics*, Spinoza defines *affectus* as: 'the affections of the body by which the power of activity is increased or diminished, assisted or checked, together with the ideas of these affections' (Spinoza 1992: 103). Shirley (the source of the cited passage here) translates this as 'emotion'; Curley as 'affect'. The latter is preferable, since Spinoza's definition does not locate *affectus* as entirely physical or mental, but rather insists on it functioning in both attributes. Deleuze and Guattari are interested in the way that affect involves a decrease or increase of power.

34. In *Francis Bacon: The Logic of Sensation*, Deleuze studies Bacon's work as an expression of this dynamic. Bacon renders visible the forces acting on flesh. See also Smith 2012: 89–106.

35. *Haecceity* dates to Duns Scotus and is often translated as 'thisness' in contrast to *quiddity* or 'whatness'. The latter refers to an essential form, whereas the former refers to particular individuality separate and distinct from the essence (Mautner 2005: 257, 512; https://plato.stanford.edu/entries/medieval-haecceity/). Deleuze's and Guattari's characterisation is in rhetorical tension with its metaphysics, since it risks reinstalling a substantialist model in referring to particles and molecules.

36. As Alliez observes, this reverses the typical order of priority following Kant: 'The question is no longer that of the *methodological* dependence of the object in relation to the subject, but of the *ontological* auto-constitution of a new subject on the basis of its objects' (2004: 56).

37. There are physiological limits as well: the ear can only register a limited bandwidth, the eye certain wavelengths, and so on. (But do we know, with certainty, how plastic these limits are, especially at their edges or perceived limits?)

38. In Chapter 6, I link this to psychologist Daniel Stern's work in what he calls 'vitality forms'.

Chapter 4

Attention, Openness and Ecological Attunement

What does it mean to say that 'individuals' are expressions of ongoing processes of individuation? How do we understand this formal innovation in relation to living experience of what is typically characterised as 'subjectivity'?

Deleuze's work, especially the co-authored texts with Guattari, more directly explores the application of metaphysics to subjectivity considered through social, political, aesthetic, linguistic and psychoanalytic registers. Deleuze is also more inclined to push normative, or better, counternormative implications of these formal metaphysics. However, while the authors of *Anti-Oedipus* and *A Thousand Plateaus* present a more fevered 'adventure' than the measured cadences of the British mathematician, the radicality of Whitehead's ideas is no less transformative.[1] Indeed, Whitehead's transitioning from the formal actual occasion to the living subject, while still abstract, exposes relevant fault lines for thinking existential implications. It is here that the connection with Deleuze, and eventually Guattari, becomes significant.

Two forms of passage and subjective attention

We have seen that individuating is a scale-relative and relational process such that basic ordering categories of 'macro'-level lived experience (the subject, the self, the object, the organism, the environment, and so on) are *abstractions* relative to occasions or events of individuating processes in question. Translating this result into existential implications is difficult since the ordinary structure of subjectivity seems to necessitate a fixed scale for at least one pole. An intuitive approach might emphasise metrical scalability, zooming in or zooming out from the sub-atomic to the cosmic. The problem with this as a heuristic is that it subtly encourages a sense that there is something fixed beneath this zooming, in this case a metric scale. For the full radicality of a process conception of individuation, the extensive metric scale cannot be presupposed as fixed prior to events of individuation.[2] This is also the case with Deleuze and Guattari's distinction between the 'molar' and the 'molecular', where the contrast is not indexed to a fixed size but is itself relative: 'the issue is that the molar and the molecular are distinguished not by size, scale, or dimension *but by the nature of the system of reference envisioned*' (ATP: 217, emphasis added).

It is one thing to think this distinction, it is another to live it or put it into practice. We can conceptually understand that adopting different frames leads to different individuations. The physicist, the biochemist, the biologist in the field frame the events constitutive of organism-environment interactions with different units of relevance and different semantics of explanation. We can understand this, even as we must conduct our own practices within framings that we only partially choose. Increasingly, we may think of ourselves as an environment or ecology, where, in a material sense, the gut's micro-biotic denizens – themselves 'independent' agential organisms – are intimately at once part of the living experience of the self or subject at another scale.[3] The question is learning to not just conceptually 'know' this, but to have this awareness impact our choices at an existential level.

Part of the difficulty is that perceptual coherence requires a degree of permanence in scale. We cannot simultaneously perceive or attend to differing scales (whether temporal or spatial) at once. If, as Whitehead reminds us: 'In all discussions of nature we must remember the differences of scale, and in particular the differences of time-span', it is nevertheless the case that existentially we privilege 'our' scale as most important (MT: 141). But the question has to do with an additional habit of fixing one scale as absolute or permanent: 'We are apt to take modes of observable functioning of the human body as setting an absolute scale' (MT: 141). This is, as Whitehead puts it, 'extremely rash' because it means 'extending conclusions derived from observation far beyond the scale of magnitude to which observation was confined' (MT: 141).

What are we to do with this? Though processes of individuation are ongoing and to some extent open, there remains a sense that self and environment are established, even static, features of life. The challenge is understanding how to destabilise this sense of static givenness *by degrees*, not categorically. Whitehead cites the famous hymn: 'Abide with me; Fast falls the eventide' (PR: 209) for how it expresses a primal tension between two ideals, permanence and flux: 'In the inescapable flux, there is something that abides; in the overwhelming permanence, there is an element that escapes into flux' (PR: 338). This tension 'formulate[s] the complete problem of metaphysics' (PR: 209).

Such a problem is not only a conceptual puzzle, it is a lived, often excruciating and painful, reality. The impact of a processual conception of subjectivity is not in somehow inoculating against the lived pain of this tension, nor in flipping conceptual privilege from permanence to flux, but rather in providing a more sustained investigation into how we understand what is permanent and what is in flux. Metaphysics are not necessary for feeling a primary tension between permanence and flux, but these metaphysics problematise the assumption that the location of permanence and the location of flux is obvious or self-evident. In a sense, they cannot be pulled apart. Instead, their entanglement is perpetually manifest in each moment's actual occasion. If it is true that we do not wake up each morning able, or having to, invent ourselves from scratch, it is also the case that who we are is not entirely given. Living the cumulative consequences of past choices and actions show that this process of individuating remains ongoing.

Understanding living subjectivity as emergent out of processes of relation inspires a different attitude of attention towards qualitative particularities because these qualitative particularities reverberate beyond the confines of any strictly demarcated 'individual' at the macro level. This claim is subject to further differentiation insofar as attention becomes part of the processes of actual occasions. How do we understand the relationship between ontological processes which are not phenomenological and attention of the living subject? On this point, Isabelle Stengers's (re) deployment of the Greek notion of *pharmakon* (φάρμακον) is pertinent:

> What characterizes the *pharmakon* is at the same time its efficacy and its absence of identity. Depending on dose and use, it can be both a poison and a remedy . . . the instability of the *pharmakon* has been used again and again to condemn it. [Against it] what has been privileged again and again is what presents, or seems to present, the guarantees of a stable identity, which allows the question of the appropriate attention, the learning of doses and the manner of preparation, to be done away with. (2015: 100)[4]

A process metaphysical conception of subjectivity intersects with attitudes of attention like a *pharmakon*. It destabilises the certainty of a fixed identity impervious to the variability of perspectives, but not as a single one-step dissolution. This point cannot be overstressed: undoing the substantial unity of a subject is not a magic bullet that automatically leads to more ecological attunement. If anything, a processual conception of subjectivity intensifies risks and difficulties inherent in social and psychic fields of interaction. Stengers's call for an art of attention is because the stakes of attention are heightened. We cannot presume stability anywhere, and yet a certain measure of stability is desired. For this reason, 'the question of appropriate attention' and its proper dose requires scrutiny into manners of understanding experiential stability as relative rather than substantial or essential.

Relative stability occurs across two different levels of transition: one is the passage of nature in which entities, selves and objects endure adventures in space and time, the second is the constitutive individuating processes operative as conditions of these entities.[5] If the former transition is more immediately accessible to lived attitudes, the entanglement of permanence and flux is enacted in these living transitions. In each actual occasion's uniqueness, the very achievement of that occasion is its perishing into an 'objective immortality' that is permanent. And yet, this closure into an 'object' is at once reappropriated by the future, which Whitehead calls the 'appropriation of the dead by the living' (PR: xiii). You live, in this sense, on the edge of a process that is always passing. Living is a manner of passage, not a stable quality. This entails a necessary openness.

While the actual occasion passes into objective immortality, it is not necessarily completely explicated. Subsequent actual occasions have the potential to express their inheritance in a manner that causes reorganising of how previous actual occasions are interpreted (or 'prehended') into the future. If every actual occasion 'just is what it is', this 'is-ness' can be differently articulated by future achievements of unification. This is both a very intuitive and very challenging point. On the one hand, it is an obvious experiential reality. We have experiences like this all the time, where we understand or 'see' the past in a different light. On the other hand, it is difficult to take seriously the real metaphysical implications here. We want to make this experience into a condition of subjective finitude, a form of our lack. To say that the meaning was there all along is to say that nothing new has really happened, we have just grown in our ability to understand a past that remains fixed. But Whitehead is suggesting that, in expressing the intersection of the potentiality of eternal forms and past occasions differently, something new has been created. In this sense, the past's potentiality is being expressed in a different way. This is only possible because the structure of the actual occasion includes all of the previous actual occasions and yet is not reducible to them. Indeed, this is what his 'epochal theory of time' amounts to, in contrast to the 'beads on a string' model.[6] To say that each actual occasion includes all of the previous is to think of it as a pulse of growth similar for example to a tree ring or a series of concentric circles.

We cannot say 'the' present, but only 'a' present. But it can be said that a present is always constituted by actual occasions that exhibit *some degree* of creation or potential for novelty. At the same time, this novelty and creation is not absolute because it occurs always in the context of what has happened already. Whitehead's description of the actual occasion as 'subject-superject' clearly exhibits these two requirements. On the one hand, as subject the actual occasion is an achievement of unification that is unique and hence logically new and ontologically creative *to some extent*. This extent varies depending on the circumstances, context and history manifest in the occasion in question. For the actual occasions active in a society characterised as a rock, their relevant creation tends, under ordinary circumstances, towards nothing.[7] As superject, however, the actual occasion exceeds its singular context in going forward as 'object' for future individuations (PR: 29). In this sense it becomes part of the inheritance that provides conditions and constraints for future 'subjective' occasions.

In this way Whitehead splits causality into *two different forms*. There is a 'self-determining' causality at the subjective level of the actual occasion, where 'feeling' is the prime operator and where a variable degree of openness remains *in principle*. Whitehead names this 'teleological' (PR: 214) where the teleology is referential to the intensity of subjective attainment in the occasion itself. And then there is the level of efficient causality from the 'super-jective' perspective of the occasion – the occasion as object for other occasions:

> there are two species of process, macroscopic process and microscopic process. The macroscopic process is the transition from attained actuality to actuality in attainment; while the microscopic process is the conversion of conditions which are merely real into determinate actuality. (PR: 214)

This metaphysical distinction is also *felt* existentially in the tension between the sense of contingency and openness that goes with living subjectivity, in which our choices matter and they feel in some limited sense up to us. And yet simultaneously, as the existentialists emphasise so well, these choices manifest in contexts we did not choose, into which we are 'thrown', as Heidegger would say. The former corresponds to Whitehead's self-determining causality at the micro level, the latter to the efficient causation excessive to individual actual occasions.

These two forms of transition are constitutive of a reality that is dynamic and creative in principle. The passage at the level of efficient causation is 'from the "actual" to the "merely real"', while at the level of the actual occasion it represents the '*growth* from the real to the actual' (PR: 214, my emphasis). A moment of present consciousness is inherently actual in its presence and yet immediately passes to the merely real, to the *what has been but is no longer*. It is now 'merely' real and no longer actual. But this reality still conditions and constrains, *but does not fully determine*, the present actual. Importantly, the *present actual* is also a *process of growth* (hence dynamic and creative). It adds to the real, it adds to what has been. This second move is *not* phenomenological. We do not likely feel as if a present moment of subjective consciousness is a creative achievement or addition to the universe. All the more so do we not think the enduring object in front of us as also constituted through micro-teleological processes of satisfaction, and yet, for Whitehead, all of reality is constituted through these two forms of transition.

This is not to say that the actual occasion is synonymous with conventional representations of lived events. Speaking technically, such representations describe occurrences at the level of what Whitehead calls 'societies': 'the real actual things that endure are all societies. They are not actual occasions . . . [because] an actual occasion has no such history. It never changes. It only becomes and perishes' (AI: 204). But this is also not to say that actual occasions bear no relation whatsoever to lived events. Rather actual occasions are the technical construction that follow a general strategy already formulated in *The Concept of Nature*.

As per that strategy, events are primary, and appeals to enduring permanence are abstractions from events. For example, writing of Cleopatra's Needle in London, Whitehead states:

> Amidst the structure of events which form the medium within which the daily life of Londoners is passed we know how to identify a certain stream of events which maintain permanence of character, namely the character of being the situations of Cleopatra's Needle. Day by day and hour by hour we can find a certain chunk in the transitory life of nature and of that chunk we say, 'There is Cleopatra's Needle'. (CN: 106)

This kind of proposition ('There is Cleopatra's Needle': a statement of location) is compared with one more traditionally describing an event ('Yesterday a man was run over on the Chelsea Embankment'), and one describing a regularity or law ('There are dark lines in the solar spectrum') (CN: 106–7). Each proposition describes different ways of abstracting from events such that 'the concrete facts of nature are events exhibiting a certain structure in their mutual relations and certain characters of their own' (CN: 107).

Though the actual occasion is a technical construction rather than a phenomenological description, it is constructed in the context of explaining concrete experience, albeit in a radically different way. This raises a question. On the one hand, we must insist that ordinary intentional representations are not synonymous with actual occasions.[8] This would suggest that conscious acts cannot directly alter the operations of occasions. Nevertheless, it is also the case that there cannot be an *essential* gap

between actions and choices of an existential subject and the actual occasions that are their constituent reality. Though not directly equivalent to
actual occasions, these actions and choices must, *in some relative way*, be
part of the inheritance of the settled past that go into the conditions of
subsequent occasions.

It seems that choices of the existential subject are, in some way, *part
of the passage of nature* that continues to inflect future creation. Here
Whitehead's tendency to move between formal technical phraseology and
applications easily enlisted into a more existential register creates a significant interpretive challenge. Whitehead says, for example, that 'it is
by reason of the constitution of the present subject that the future will
embody the present subject and will reenact its patterns of activity' (AI:
193). Any Whitehead scholar will likely remind you that he is not talking
about human subjects here, correct? The subject is the actual occasion,
not the society, and the human subject is a society. And yet, how could
the choices of the society not bear on its constitutive actual occasions?
Moreover, this remark is made in *Adventures of Ideas*, not *Process and
Reality*, a text in which Whitehead develops a reading of history through
his metaphysics.[9]

If these choices are part of the passage of nature, there remains the question of *whose choices they are*. If acts of attention are part of the conditions
of creation that inform individuated occasions, we should not automatically enlist this into a view of the autonomous subject wilfully directing
these acts by caveat. The situation is much stranger than that. Rather than
essentially ephemeral or merely transitive, occasions *do not simply disappear*, but are part of the objective datum that go forward into the universe
as both constraint and possibility for the future. This is to say that there are
no *purely* neutral or passive bare receptions of what is simply there – *each
act of attention is an act of partial creation* though no creation is *ex nihilo*.

Subject as society: openness and stability

Details of Whitehead's conception of 'society' are necessary to develop
this claim about attention. Whitehead defines society as 'a nexus with
social order' (PR: 34). This definition is circular. Though in a loose sense a
society is a grouping of occasions that share certain qualitative features in
common, we must track the painstaking way in which Whitehead builds
this concept to understand its existential stakes without assimilating these
into prevailing assumptions. This is also necessary because he is careful to
stress the *difference* between metaphysical categories and empirical concepts drawn from the physical and biological sciences, even as mapping

metaphysical categories to the empirical is necessary to meet the criteria of 'adequacy'.[10] Though he offers a typology of societies characterised as a 'hierarchy' with a nested, Russian-doll-like structure, and lists as 'general examples' individual electrons, protons, individual molecules, societies of molecules such as inorganic bodies, living cells and societies of cells such as animal and vegetable bodies (PR: 96–8), we must begin with the metaphysical construction, not the empirical examples.[11] The persistence of this challenge cannot be underestimated: what Whitehead's society collects are kinds of occurrences – not things, not objects, not entities. It is an entirely different way of describing *the genesis* of taxonomies precisely because it rejects the givenness of the 'things' being grouped.

Understanding the circularity of the definition in a productive rather than vicious sense requires distinguishing two different methods of specifying a 'set' of occasions.[12] A nexus is 'a set of actual entities . . . constituted by their objectifications in each other' (PR: 24). Given the relational structure of actual occasions and operations of negative prehension, Whitehead's definition here does not determine any particular group. Instead, it marks the possibility of grouping occasions relative to any chosen feature or relation whatsoever. Think of Foucault's reference to Borges's 'Chinese encyclopaedia' at the outset of *Le Mots et le Choses*, where the appearance of alternative categories exposes the contingency of taxonomy and possibility of other groupings: 'animals divided into: (a) belonging to the Emperor, (b) embalmed, (c) tame, (d) sucking pigs, (e) sirens, (f) fabulous', and so on (1994 [1966]: xv). Such spirited questioning of the *production* of order is endemic to the role of speculative philosophy in questioning what is presumed unquestionable. And yet this questioning is not conducted by Whitehead (or for that matter Deleuze) as end-in-itself. Nor are their efforts content to exclusively historicise or periodise or culturise. The germ of the metaphysical remains in seeking to understand the process of the production of orders without presuming the stability of any particular empirical order relative to any 'cosmic epoch'.

This helps explain Whitehead's neologisms, which emphasise the novelty of his effort in departing from assumptions of given orders while nevertheless developing conceptual tools to understand how order as such is generated. A nexus can then be thought of as the first, least differentiated such tool. A nexus need not reflect any internally determined order and can be entirely imposed from a perspective external to the nexus in question. It can reflect any happenstance or arbitrarily chosen collection. By contrast, a society is a nexus that 'enjoy[s] social order' where this means an internally generated coherence. Whitehead describes this through three technical criteria:

> (i) there is a common element of form illustrated in the definiteness of each of its included actual entities, and (ii) this common element . . . arises in each member . . . by reason of conditions imposed upon it by its prehension of some other members of the nexus, and (iii) these prehensions impose that condition of reproduction by reason of their inclusion of positive feelings of that common form. (PR: 34)

Social order requires a mutually sustaining relationship between occasions that is positive. It is not based on what the occasions leave out (negatively prehend) but rather what they feel in relation to a form included in the self-satisfaction of each occasion. Importantly, this feeling is both in relation to eternal objects and to each other:

> The common element of form is . . . a complex eternal object exemplified in each member of the nexus. But the social order of the nexus *is not the mere fact of this common form exhibited by all its members.* The reproduction of the common form throughout the nexus is due *to the genetic relations of the members of the nexus among each other, and to the additional fact that genetic relations include feelings of the common form.* (PR: 34, emphasis added)

This is significant because it shows the extent to which a society is self-defining through its manner of feeling potentials (eternal forms) that exceed its own perspective. As Whitehead puts it, 'the point of a "society" . . . is that it is self-sustaining . . . *it is its own reason*' (PR: 89, emphasis added).[13] It also shows the extent to which this feeling of forms can vary within the society and that the society's occasions impact the potential for how other 'members' of the society feel the forms.

Questioning the construction of order does not make it any the less real. Indeed, Whitehead's concept of 'social order' is a feature of his dynamic realism. Because a society is its own reason, the concept responds to the fundamental intuition of all realisms: reality is excessive to human cognition of it. The reality of a tree does not *depend on*, in any ontological sense, its human categorisation or taxonomic description. Nevertheless, it is also crucial that Whitehead's concept is, in a sense, content neutral. While empirical examples are useful as explanatory heuristics, Whitehead is describing the genesis of groupings of occasions, not a static taxonomy akin to natural kinds of entities.

In this way, Whitehead is able to respond to a basic intuition of a Foucaultian approach: categories are, to some extent, relational and relative. Indeed, the complexity of Whitehead's analysis of societies goes much further in this direction even as it pushes relationality beyond exclusive reference to human epistemes or subjectivities. The 'social order' of societies is a metaphysical concept following the relativity of all occasions. Any concrete example must be taken with care since it comes with a temptation towards reification that would fix the content of a social ordering at the expense of Whitehead's real goal, providing a means for thinking about societies as orderings in process, not groupings of static entities.

This has a number of consequences with regard to what Whitehead calls societies with 'personal order' (PR: 34).[14] Such a society 'orders its members [actual occasions] "serially"' (PR: 34). This means that they have a history that informs their nature. Whitehead presents this in a highly technical fashion to trouble the desire to easily transcribe it into conventional modes:

> any member of the nexus – excluding the first and the last, if there be such – constitutes a 'cut' in the nexus, so that (a) this member inherits from all members on one side of the cut, and from no members on the other side of the cut. (PR: 34)

This accounts for a serial linearity of perduring entities without presuming this linearity to be a pervasive feature of reality as such. He thus opens the possibility for other forms of societies or nexus whose serial order is of a different duration.

A serial ordering means that the society 'sustain[s] a character' that nevertheless is not static. An enduring entity can maintain a recognisable coherence and yet change over time. It is these societies which present the 'enduring objects' that 'enjoy adventures of change throughout time and space' (PR: 35). Importantly, the category of 'personal order' or 'enduring object' does not necessarily imply life, but rather recognisable persistence. Nevertheless, living subjects, organisms and human persons are included as more specialised versions of such societies. Whitehead writes, 'the life of man is a historic route of actual occasions which in a marked degree . . . inherit from each other' (PR: 89). After making this declaration, Whitehead offers an odd example:

> That set of occasions, dating from his first acquirement of the Greek language and including all those occasions up to his loss of any adequate knowledge of that language, constitutes a society in reference to knowledge of the Greek language. Such knowledge is a common characteristic inherited from occasion to occasion along the historic route. (PR: 89–90)

The peculiarity of the example is telling, and he qualifies it with a claim that 'this example has purposely been chosen for its reference to a somewhat trivial element of order, viz. knowledge of the Greek language'; before a casual admission that he is dodging the weightier question of personal identity: 'a more important character of order would have been that complex character in virtue of which a man is considered to be the same enduring person from birth to death' (PR: 90).

Whitehead's peculiar example along with his seemingly casual evasion hints at profound implications. In choosing the 'trivial' element of knowledge of Greek, Whitehead shows the inherent relationality of social order and hints at pertinent existential difficulties. There would appear to be no singular definition that can fix a living subject as one society alone. Rather, such a subject can be relationally described through different privileging of perspectives: i.e., a particular aptitude or knowledge, one's 'self' as a function of their relations rather than vice versa. Societies are thus forms of order such that the 'same' macro-level object or subject may be constituted by actual occasions that also participate in other forms of order. Indeed, given their relational structure, this must be the case. Whitehead's reference to the 'complex character' of an enduring person is less a dodge than it first appears. Indeed, it is rather an acknowledgement that the 'sameness' of the enduring person is not a fixed predicate, but a complex function that is necessarily perspectival. It is the 'character *by which* a man *is considered to be the same*' not the designation of the quality that *is* the same.

The complexity of this character follows Whitehead's insistence that 'there is no society in isolation' (PR: 90). Rather, 'every society must be considered with its background of a wider environment of actual entities, which also contribute their objectifications to which the members of the society must conform' (PR: 90). As collections of actual occasions 'imposing' on each other conditions leading to a common character, societies cannot be extracted from wider environments of actual occasions that do not share this character. Their ability to persevere requires negotiation with wider conditions. The reason for a society's endurance cannot appeal to an essential form of the society in question. Since actual occasions are always the basis for reality, relative persistence within reality must be explained through actual occasions, not imposed on them.

Societies exhibit different degrees of stability, complexity and relative independence.[15] Some barely last more than a few moments, others (sequoia trees, for example) prevail for several hundred years or longer. Some are uniform and simple, while others are highly complex and heterogeneous. Stability is a function of how the society negotiates its relations with the wider environment:

> A society is 'stabilized' in reference to a species of change when it can persist through an environment whose relevant parts exhibit that sort of change. If the society would cease to persist through an environment with that sort of heterogeneity, then the society is in that respect 'unstable.' A complex society which is stable provided that the environment exhibits certain features is said to be specialized in respect to those features. (PR: 100)

Highly specialised societies are likely to be unstable in conditions of change. Because of their specialisation, they have a narrow set of parameters for survival. But, the degree of 'satisfaction' of the members of a society (actual occasions) is more intense the greater the complexity of contrast involved in the achievement of each occasion. A complex society achieves a more intense reality. So: 'the problem for Nature is the production of societies which are "structured" with a high "complexity" and which are at the same time "unspecialized." In this way, intensity is mated with survival' (PR: 101).

There are two different tendencies by which a society can achieve intensity with survival: '[T]he two ways in which dominant members of structured societies secure stability amid environmental novelties are (i) elimination of diversities of detail, and (ii) origination of novelties of conceptual reaction' (PR: 102). The first involves 'eliciting a massive average objectification of a nexus, while eliminating the detailed diversities of the various members of the nexus' (PR: 101). This means reducing and simplifying as much as possible to 'block out all unwelcome data' (PR: 101). In effect, the society endures through consolidating potential variety so that it is 'supported by a massive objectification of the many environmental nexus' (PR:101). This is the solution achieved by structured societies that are standardly classified as inorganic bodies ('crystals, rocks, planets, and suns') and is enormously successful: 'Such bodies are easily the most long-lived of the structured societies known to us, capable of being traced through their individual life-histories' (PR: 102).

The second tendency solves the problem in a contrasting way. Instead of achieving an intensity of persistence through narrowing the relevant datum through which the society achieves its character, it involves 'an initiative in conceptual prehensions, i.e., in appetition' (PR: 102). Rather than denying the relevance of novel elements of the environment through the procedure Whitehead calls 'negative prehension', such novel elements are 'received' by 'explicit feelings with such subjective forms as conciliate them with the complex experiences proper to the members of the structured society' (PR: 102). The subjective aim of the individual occasions 'originates novelty to match the novelty of the environment' (PR: 102). Such a potential is inherent in actual occasions, but at the level of the society it manifests as a capacity to receive and respond to novel elements and changing conditions. Whitehead notes that 'structured societies in which the second mode of solution has importance are termed "living"' (PR: 102).

This second choice is not just a matter of being passively flexible, but also of being actively creative in inventing new responses. Though Whitehead notes, 'In the case of higher organisms, this conceptual initiative amounts to *thinking* about diverse experiences' (PR: 102), the abstraction of his description should be maintained.[16] Thinking understood in the sense of human consciousness is not required for this solution. It is a matter of how occasions achieve satisfaction and recognisable conformity to establish the society's order. This tendency need not be equally manifested in every occasion constituting the society in question – especially since the society in the case of an organism is constituted through complex organisations of materiality that also involve the first solution. There is no categorical hierarchy for understanding how these tendencies and solutions relate in a complex structured society. As always, the variance is by degrees.

As a heuristic, this metaphysical understanding enables differing interpretations of organic behaviour. Massumi for example draws on this Whiteheadean framework to argue that so-called 'instinctive' behaviours are not opposed to the creative, but are rather 'inextricably entwined' with it (2014: 91, 2015). Rather than deterministic binding structures, instinct responses are modes whereby the organism seeks new expressions of fundamental appetitions. The range of variability within this expression is wider than a direct stimulus-response à la Pavlov. Working from ethologist Nikolaas Tinbergen's studies in herring gull chick's recognition patterns, Massumi observes that instinctive activity has a tendency towards what Tinbergen calls 'supernormal stimuli' and declares that 'the force of the supernormal is a positive force' (2015: 9). That is, where mechanical force pushes, supernormal force pulls or attracts. In this sense, supernormal stimuli are structured as means by which behaviour can exceed or disrupt the conformity of the repeated past. That is, they are attractors which enable the organism to navigate creatively.[17] This is precisely the possibility of novelty within a restricted range that Whitehead enables.

Whitehead's metaphysics can also be applied to neurological discourses of brain plasticity. As Catherine Malabou has argued, a challenge of 'plasticity' involves articulating its concept in a way that does not reductively function as 'unconscious justification of flexibility without limits' (2008: 13).[18] A proper concept of plasticity emphasises not such unlimited flexibility, but rather the reality of repeated contingent choices in building the brain: 'everything begins with establishing connections and then multiplying them and making them more complex' (2008: 18). Developmentally, this begins below the level of conscious intentionality, but the process continues beyond early development. As Malabou puts it, 'humans make their own brains, and they do not know that they do so' (2008: 12). The important distinction is between developmental plasticity and modulational plasticity. Whitehead's distinction between open and closed forms of perseverance enables an understanding of modulational plasticity defined as 'the modification of neuronal connections by means of the modulation of synaptic efficacy' (2008: 21). This form of plasticity refers especially to the extent to which surroundings and the choices enabled by those surroundings create (or do not create) neural pathways. The potential openness of the brain is of course always modulated in specific and contingent ways. As Malabou puts it, modulational plasticity shows that 'there is a sort of neuronal creativity that depends on nothing but the individual's experience, his life, and his interactions with surroundings' (2008: 21–2). As a society, the brain thus displays essential features of the Whiteheadean 'open' choice for perseverance. However, and crucially, this is also to say that each act of attention and action is potentially implicated in the very possibilities of future attention and action. This is not a deterministic causality, but rather a cumulative one. Each achievement of satisfaction of an actual occasion constitutive of the 'brain society' structures possibilities going forward, but never in a completely deterministic manner.

The distinction between these solutions (open and closed) is thus not mutually exclusive. Even as the brain's manner of complexity is defined by a tendency towards incorporating novelty (whether in 'modulational plasticity' or, when necessary, in 'reparative plasticity'), this complexity is shaped by anatomic or physiological form (echoes of Spinoza's famous question about the limits of the body). Moreover, the difference between Uexküll's infamous tick and a complex mammal is significant, though both are considered living. Among living societies, there is the distinction between vegetable and animal solutions as well, between those which develop a prevailing 'centre of experience' (the brain) and those 'which decisively lack any one centre of experience' (MT: 24).[19] For this reason:

> It is obvious that a structured society may have more or less 'life,' and that there is no absolute gap between 'living' and 'non-living' societies ... A 'living society' is one which includes some 'living occasions.' Thus a society may be more or less 'living' according to the prevalence in it of living occasions. (PR: 102)[20]

Occasions also follow this variability: 'an occasion may be more or less living according to the relative importance of the novel factors in its final satisfaction' (PR: 102). Life is not a static predicate, but a manner of *how* occasions achieve satisfactions and *how* societies manifest responses to changing conditions. These responses are inherently relational, and the characterisation of 'living' or 'life' thus marks a feature or style of passage between actual occasions. As Whitehead puts it: 'life lurks in the interstices of each living cell, and in the interstices of the brain' (PR: 105; Debaise 2013).

This responds to the multi-scalar complexity of occasions that negotiate differing series of encounters, thresholds and transitions across membranes, skins, territories and borders. Such encounters are constitutive of the relational processes driving individuation, what Keith Ansell Pearson describes as the 'individuating closure' that all life requires. As per Chapter 2, the crucial feature is that these membranes, skins, borders and surfaces bring into communication two disparate levels or interiors and exteriors that are scale-relative (Ansell Pearson 1999: 210). This multiplies the individuation processes collectively operative in various forms of social order: 'An integral living society, as we know it, not only includes the subservient inorganic apparatus, but *also includes many living nexūs*, at least one for each "cell"' (PR: 103, emphasis added). While common sense is inclined to see the organism as constitutive of the distinction between inner and outer (the organism exists, and then there is an interior and an exterior relative to it), for Deleuze and Whitehead the existence of the organism is an expression of a communication between disparate orders. A society-organism emerges out of this communication, which occurs through series of insides and outsides negotiated at different levels and scales.

The difference between life and death is akin to diverging modes of relation where one mode practises an art of novel response and the other an imposition of conformity. These terms are not axiological or normative, but rather descriptive. Both 'solutions' respond to the metaphysical problem of persistence given the premise of change (in Deleuze's language 'difference') as ontologically primary. Whitehead thus criticises the tendency of popular evolutionary metaphors for committing the 'fallacy . . . [in believing] that fitness for survival is identical with the best exemplification of the Art of Life' (FR: 4). Fitness for persistence cannot explain why complex organisms, which are enormously more sensitive and hence precarious and vulnerable to environmental change, would have evolved.

Whitehead's metaphysical orientation offers an approach to the question of survival in the abstract as one of persistence. That is, he does not beg the question by first presuming 'life' as the only relevant class for thinking persistence. As he puts it in characteristic droll fashion:

> the art of persistence is to be dead. Only inorganic things persist for great lengths of time. A rock survives for eight hundred million years; whereas the limit for a tree is about a thousand years; for a man or elephant about fifty or one hundred years, for a dog about twelve years, for an insect about one year. (FR: 4–5)

What Whitehead calls 'the Art of Life' therefore cannot be assimilated easily into a question of evolutionary fitness where fitness reduces to persistence. The problem changes form, since 'life' is a relational feature of transition. Transition happens, this is fundamental. There is no evading it. The question has to do with how a given society negotiates such transitions so as to achieve the coherence of a sustaining persistence. This is an abstract question faced by planets, crystals, bacteria, flowers, rhizomatic grasses, oak trees, dogs and human beings.

The experience of societies varies according to how they negotiate this question. For a society oriented towards openness, qualitative experience will be intense in its variability and contrasts. For a society oriented towards closure, intensity is achieved alternatively through the omission of relevant difference. Rather than an intensity of contrast, it is an intensity of conformity. Persistence in this case involves less qualitative variability. Neither solution is categorical. Every society must continually 'solve' or negotiate its degree of openness.

For purposes of example, consider a boulder that has been laid on a forest floor by a retreating ice sheet many centuries ago as a society that approaches closure (keeping in mind that the relative endurance of a society depends on the time-scale in question and that 'complete' closure or openness are limit conditions). We can emphasise the contrast between this boulder as a society approaching maximal closure and an organism such as a deer or wolf that lives in the same forest and is relatively more open. Whereas such an organism is relatively susceptible to changes in temperature, light and the presence of other forms of life as either predator or food source, the boulder has a much wider range of non-sensitivity to changing conditions. It does not require food. It is immobile. It does not require light. Swings in temperature and moisture largely do not effect it. (This does not mean that it is completely insensitive. At certain extreme high temperatures or low temperatures, the integrity of the boulder as society will break down.) From the temporal and physical scale perspectives of a human being, the boulder, as a relatively closed society, is much more impervious to change than other societies in the same milieu. But this is paid for at the cost of a loss of flexibility (mobility is the obvious example) and *less potential variety in range of experience*. The contrasts of its occasions occupy a much narrower ranger, with the benefit of an increased power of persistence over change.

It cannot be stressed enough that these two tendencies are not mutually exclusive and that all societies negotiate their response to the problem of stability, persistence and intensity of experience along a continuum that partakes of different degrees of both. Nevertheless, at the risk of overly anthropomorphised projection, we can make some cautious translations into existential language. The closed society can be characterised as a *becoming-zombie*, exemplified in addictions or fixations of various forms: shut out as much as possible, fix on what 'works', achieve stasis. The other extreme of an almost entirely open society would be highly volatile and unstable. Given the dynamism of occasions, this extreme tends towards dissolution. While it may achieve intense degrees of experience, its persistence is likely to be short-lived.

If any individuated society must negotiate this tension between closure and openness, societies that are characterised as living face a much greater dynamism in such negotiation. *It is too simple to say that life is open and death is closed.* All living societies require some degree of closure to maintain integrity. But at some point this tendency towards closure tips the threshold when the overriding strategy of perseverance becomes imposition of a narrower range of relevant inheritance in occasions. A society is no longer living when its overriding strategy of persistence is to repeat the same form or element of experience and reduce, deny or dismiss alternative elements. Again, this is always a question of degrees and both tendencies come with risks and benefits. Orienting towards closure has the advantage of resulting in predictable outcomes (because 'data' that do not fit the outcome are denied or selectively ignored), whereas the other has the danger of collapse into incoherence. But, simultaneously, the intensity of satisfactions of occasions within a society tending towards death will tend towards diminishing return and prevailing dullness. This results in an increasing lack of sensory discernment and an increasing homogenisation of relevant surroundings, so that one becomes a data processing machine fixed to one loop, a drone, a capitalist consumption robot. Tending in the direction towards life is undoubtedly more immediately risky, if only because it requires a non-categorical stance for creating responses to conditions. It 'gains intensity through freedom' (PR: 107), but the price of freedom is the greater possibility of self-disintegration.

Whitehead's presentation of two choices with regard to the metaphysical problem of persistence firmly locates this problem – whether in reference to an organism, an institution or a subjective individual – in the context of ecological relation. How an individual persists (or does not persist) is a function of its capacity and manner of responding to environmental change and variability. While degrees of this capacity vary (historically, geographically, topographically) Whitehead's two basic strategies provide a basis for thinking of the stakes of an ecological attunement in relation to practices of attention. Is attention oriented towards dismissal of variety in the interests of maintaining predictable stasis or towards the cultivation of creative response? The latter choice raises challenging questions, since, as Whitehead puts it, such creative response 'may be unfortunate or inadequate . . . [or even amount to] persistent failure' (PR: 102). There is no guarantee that creative response will 'work', and if persistence is taken as its only criteria for success, then inevitable conflicts between differing forms of living response seem likely.

For this reason, the respective balance struck between the asymptotic extremes of maximal openness or complete closure (both extremes being *socially* impossible) cannot be prescriptively legislated in a universal manner. How this balance is (or is not) achieved depends on context and the individual in emergence, and on respective affordances, capacities and 'environmental' conditions. The analysis here is resolutely metaphysical. Whitehead does not begin with *any essential difference in kind between classes of entities*. The problem 'for Nature' encapsulates the range of societies from the molecular to the cosmic, from crystals to planets, from the micro-biotic to apex fauna. This is also to say that respective solutions to this problem of persistence are descriptive only. Whitehead is not proposing normative valence to various solutions, especially at the formal level. Is one form of persistence 'better' than another (is a tree 'better' than a river?). There is no privileged exception for the human.

Nevertheless, his analysis lays out fault lines of a persevering individual that are not without implications in more (humanly) existential or subjective registers. In this sense, Deleuze and Guattari's deployment of their concept of a 'line of flight' ('lignes des fuite') is best understood as operative within this Whiteheadean tension between closure and conformity and creative openness or novelty.[21] The 'line of flight' takes the second 'solution' to a more intense extreme in the context of an applied metaphysics that includes historical, political and economic forces as relevant. It is crucial for generating creative response to powerful forces (social, political and subjective) that would choose closure as a means of maintaining control. In the next section, I consider this deployment in the context of 'subjectification', paying particular attention to effects in the experience of normativity.

'Lines of flight' and paradoxes of normativity

Deleuze's emphasis on experimentation responds to the potential hegemony of Whitehead's first solution in human contexts. Deleuze and Guattari's 'lines of flight' function both as a descriptive metaphysical concept and, to some extent, as a prescriptive call. Descriptively, 'lines of flight' correlate with 'movements of deterritorialization' as processes that destabilise, dissolve or otherwise challenge 'molar' aggregates or assemblages. A 'line of flight' enables passages from the 'plane of organization' to the 'plane of consistency' causing the latter 'to rise to the surface' (ATP: 270).[22] The plane of consistency is characterised by events, affects and flows rather than stable molar entities. In this sense, it is akin to the reality of actual occasions prior to consolidation in macro-level societies. Lines of flight therefore play the ontological role of enabling novelty through the destabilisation of sedimented forms towards the emergence of an alternative becoming. Whereas conventional identity categories are molar, becomings involve relational exchange at the molecular level: 'all becomings are molecular: the animal, flower, or stone one becomes are molecular collectivities, haecceities, not molar subjects, objects, or forms that we know from the outside and recognize from experience, through science, or by habit' (ATP: 275).

A 'line of flight's' processes of transition can occur at different levels or 'strata'. In this sense, the concept of 'line of flight' is consistent with Whitehead in *offering no privileged exception for the human*. Lines of flight are ontologically descriptive, not normative. Deleuze and Guattari repeatedly insist that 'lines of flight' carry significant risks: 'so much caution is needed to prevent the plane of consistency from becoming a pure plane of abolition or death, to prevent the involution from turning into a regression to the undifferentiated' (ATP: 270).[23]

Though 'lines of flight' do not suffice as liberatory ends-in-themselves, they are particularly relevant to processes constitutive of 'subjectification'. Indeed, the general tension between the two solutions that Whitehead proposes are *exemplified in subjectification*. Subjectification (the *becoming-Cogito* of prevailing neo-Cartesian models of consciousness) presumes to absolutise oneself against constraints of the merely material (*res extensa*). Deleuze and Guattari observe that 'subjectification assigns the line of flight a positive degree [and] carries deterritorialization to the absolute' (ATP: 133). Becoming a subject appears as emergent transcendence that escapes strict determinism and generates consciousness as a freedom or 'line of flight' from its preceding conditions. But this escape is relative, not absolute, and carries with it a backlash or reversal: it 'has its own way of repudiating the positivity it frees' (ATP: 133) as 'subjectification imposes on the line of flight a segmentarity that is forever repudiating that line, and upon absolute deterritorialization a point of abolition that is forever blocking that deterritorialization or diverting it' (ATP: 134).

It would however be too simple to define subjectivity only through this line of flight. Though a line of flight has a privileged relation to subjectification, it is not exhaustive, since the process of subjectification entwines with two other lines: (1) 'a molar or rigid line of segmentarity' and (2) 'a line of molecular or supple segmentation' (ATP: 195–6). The first line of molar segmentarity provides basic templates for the conventional representations of life. On this line, life is a 'whole interplay of well-determined, well-planned territories' that 'have a future but no becoming' (ATP: 195). While surface correlates will change, this is nothing more than a change of scenery and there is nothing truly unexpected: life involves finding a career, a partner, and so on. Though one is tempted to caricature this line into the stale conformities of a repressed bourgeois, Deleuze and Guattari make clear that this line 'pervades our life . . . [and] includes much tenderness and love [such that] it would be too easy to say, "this is a bad line," for you find it everywhere, *and in all the other lines*' (ATP: 195, emphasis added).

The presence of the molar line in the others exemplifies the perspectival complexity of subjectification as process. The molar is real, but it is not the whole story. If its relations or segments involve 'well-determined aggregates or elements (social classes, men and women, this or that particular person)' these well-determined elements cannot be fully disentangled from the second 'molecular or supple' line of 'micromovements . . . tiny cracks and postures operating by different agencies even in the unconscious' (ATP: 196). Determinations of the molar are both constituted by and yet also challenged through micromovements of the molecular. This is a function of the dynamism of the molecular, in which a supple flow is 'marked by quanta that are like so many little segmentations-in-progress grasped at the moment of their birth' (ATP: 195). These segmentations-in-progress are akin to actual occasions in concrescence, that is, actual occasions as subjects, not determined superjects. And yet, from the molar perspective, this line is lived in the 'form of something that has already happened, however close you might be to it, since the ungraspable matter of that something is entirely molecularized, traveling at speeds beyond the ordinary thresholds of perception' (ATP: 196). It is as if we understand that we are constantly grappling with effects of a myriad of micro-perceptual encounters, and yet we always only cognise these effects after they have occurred.

Quanta are 'ungraspable' through molar representations of ordinary consciousness, but nevertheless *lived* through processes of subjectification. It cannot then be a question of isolating the 'good' line, since all three (molar/rigid, supple/molecular, line of flight as deterritorialisation of all segments) 'are constantly interfering, reacting upon each other, introducing into each other a current of suppleness or a point of rigidity' (ATP: 196). The distinction between the ordinary and the singular (Chapter 2), in the form of qualitative thresholds and tipping points, is operative within this entanglement of three lines. There is thus a constitutive tension between fixed determination and destabilising excess that goes beyond determinations in unpredictable ways. At what intensity do the micromovements of the supple transform into a line of flight reorganising the molar into a new identity?

Deleuze and Guattari's three lines stress the impossibility of isolating a sole foundation for the subjective individual. Instead, processes of subjectification proceed through the ongoing entwinements of all three lines.

Privileging any one as logically or ontologically primary is a reduction. Given their complexity and abstraction, it is tempting to minimise the risks of such reduction as primarily formal, or to dismiss them as merely metaphysical. But a prime merit of the schizo-analysis project is how it dwells at the intersection between abstract and lived – all metaphysical exploration is manifested through applications across edges of disciplines or domains, always with an eye to lived effects. If the danger of solely prioritising any one line is what it omits descriptively, each line also has a corresponding lived danger:

1. The danger of the rigid line of segmentarity is its tendency to stagnate and enforce hegemonic conceptions of identity. When the function of 'ensur[ing] and control[ling] the identity of each agency, including personal identity' (ATP: 195) is presumed metaphysically or existentially exhaustive, it is easily enlisted by forces seeking to perpetuate homogenous conceptions of life and normativity. This danger is salient in the context of IWC's (Integrated World Capitalism) operational insistence on equivalence as sole dominant rubric of value (Chapter 5). Because the rigid line denies the possibility of becoming, it enables an imposition of norms to reinforce predictability in the service of profit.

2. Given this risk of enforced stagnation, the second supple line of the 'molecular' can be seen as its counter-ballast in 'concern[ing] flows and particles' that 'elude' those of the rigidly segmented or defined 'classes, sexes and persons' (ATP: 196). But this elusion is not in-itself always 'positive'. Because the molecular cannot be accessed through ordinary modes of representation or conscious cognition, it is at risk of being ignored or denied, or, more troublingly, co-opted through rhetorical appeals to micromovements that selectively fixate on some of its effects at the expense of others. The molecular is crucial for the potential of creative becomings, but its incumbent 'disorientation' carries its own risks, especially when reterritorialised by a dominant, though different, rigid segmentation.[24]

3. The 'line of flight' exceeds the micromovements of the molecular (which still are reterritorialised within processes of subjectification). It is like 'an exploding of the two segmentary series' and a kind of 'absolute deterritoriliztion' (ATP: 197). Though Deleuze and Guattari often leverage its creative potentials, the risks here are even greater, since 'there is a danger that these vibrations traversing us may be aggravated beyond our endurance' (ATP: 197). The line of flight expresses the possibility of 'becoming imperceptible', not towards nihilistic dissolution, but so as to open subjectivity to a less mediated connection with the events through which it is constituted. In this way, we might

'make consciousness an experimentation in life' and even 'use love and consciousness to abolish subjectification' (ATP: 134). Though the abolishment of subjectification is not end-in-itself (it can just as easily result in destructive dissolutions that give rise to further rigid segmentarities and fascisms), it carries with it the potential for ontological creation as a collaboration with forces and events of the real ('the possibility of a positive absolute deterritorialization on the plane of consistency' (ATP: 134)). It is as if we have to get the fixations constitutive of the ordinary subject out of the way if we hope to encounter the intensities and events that are always already there. This is 'to have dismantled one's [molar, fixed, determinate] self *in order to become capable of loving*' (ATP: 197, emphasis added).

Though their description of subjectification is consistent with Whitehead, there is nothing precisely like the line of flight in Whitehead. In one sense, the line of flight functions like a release valve that destabilises any appeal to a final teleology or holism. Interest in such a teleology or holism, especially approached through potential theological implications of Whitehead's God, is a legacy of the Hartshornian lineage in process thought. Brian Henning for example stresses Whitehead's processual universe as teleologically ordered towards the production of beauty in the interests of developing a holistic axiology and ethics.[25] At first glance, the line of flight is directly at odds with such a project. However, from another perspective, the line of flight's renewal might be *consistent with this holism* insofar as it invokes the destabilisation of any molar entity taken as independent or separate. The 'plane of consistency' might be seen as playing a holistic role insofar as it encompasses *all of the ontological dimensions into a plane of events*. But everything depends on how we understand this holism, especially for endeavouring to find normative or axiological rudders.

For Deleuze and Guattari, extraction and representation of such rudders carries significant risks of reification. While experimentation is advanced through the ideal of manifesting the 'plane of consistency', it does not translate to fixed normative or axiological propositions.[26] Many Whitehead scholars approach this question through his tripartite depiction of value: 'everything has some value for itself, for others, *and for the whole*: this characterizes the meaning of actuality' (MT: 111, emphasis added). This introduces a pervading problem for thinking the existential implications of such metaphysics: how to construe the presence of the whole as a felt element in living occasions? While it is undoubtedly right formally to say, with Henning, that Whitehead's ethical imperative is to 'maximize the beauty, value, and importance possible in each situation' with an aim of 'affirm[ing] the most beautiful whole possible', it remains unclear how any particular society lives access to the (theoretical) whole (Henning 2005: 171–2). Is it possible to make claims from the perspective of the whole? Indeed, any concrete normative appeal to a valuation that claims to leverage the perspective of the whole is fraught with historical difficulty given how often such appeals have functioned as apologies for colonial domination, dismissal or eradication.[27]

It is true that Whitehead unequivocally characterises beauty in teleological terms. However, when Whitehead says that 'beauty is the one aim which by its very nature is self-justifying' (AI: 266), the locus of this self-justification, as we have seen, is the occasion realising itself and its 'certain absoluteness of self-enjoyment' (MT: 150). *This beauty, in itself, has no aim other than its own realisation.* So, while a realisation of beauty guides the process of unification in each actual occasion, this realisation is local and self-referential and does not necessarily refer to an overarching telos of the universe as such. And yet, it does add to this universe. Its realisation contributes to the universe, but it does not necessarily lead in a direction of greater beauty. This is why evil is a necessary feature of actuality since there can be no purely conceptual way of guaranteeing the commensurability of the subjective aim of all actual occasions. As Whitehead writes in *Religion in the Making*, 'there is evil when things are at cross purposes' (RM: 97). By *Process and Reality*, this observation has shifted to become unavoidable, not only is there evil when things are at cross-purposes, but things *will be* at cross-purposes: 'The nature of evil is that the characters of things are mutually obstructive' (PR: 340).

It is undoubtedly correct that Whitehead 'leans ethics upon aesthetics' and that this orientation is compatible with Deleuze and Guattari. It is also correct that the formal criteria for a choice must be in terms of what will realise the greatest potential for beauty as a whole. Ideally, this is a mutually resonant situation in which an individual satisfaction contributes to greater possibilities of satisfaction beyond. But we must remain wary of appeals to beauty which cover over the existential challenges of perspectivism. There is a particular danger in setting up an appeal to the whole if we think it translates directly or easily into an existential perspective. This challenge is salient in considering the intersection of three differently valenced criteria: beauty, harmony and intensity. Henning presents these as mutually reinforcing such that 'for Whitehead what is preferable is a state of affairs that achieves the most harmonious and intense beauty in the situation *taken as a whole*' (2005: 112, emphasis added). But there is no guarantee that the intensity of the beauty, taken as a whole, simply passes down unchanged into the perspective of a given society or individual. The whole is just what we cannot know. Moreover, remaining attuned to the intersection of theory and living effects means being mindful of the extent to which living individuals can be defined, in part, through negative prehensions. Any existential society functions in part through its manner of abstracting from the whole *where this means an omission or reduction*. Appeals to holistic beauty therefore do not map onto the lived experience of a social individual in any way that can be easily accessed without remainder.

This tension is why Deleuze identifies what he calls the 'beautiful soul' as the 'greatest danger' for a processual philosophy of difference (DR: xx). The beautiful soul 'sees differences everywhere and appeals to them only as respectable, reconcilable or federative' therefore 'behave[ing] like a justice of the peace thrown on to a field of battle' (DR: 52).[28] In effect, the beautiful soul assumes that harmony=intensity=beauty necessarily. But even in Whitehead this is not the case, since intensity can be generated both through the imposition of conformity (Whitehead's first solution) and through a pushing of experience to the limits of any molar identification. Beauty as an ideal cannot be construed as a static harmony because the dynamism of a processual view means that the quality of living experience cannot be frozen and judged from the perspective of a final harmony. God in Whitehead is never such a judge. For this reason, Hartshorne's 'Diagram of aesthetic value' does not directly engage the existential question.[29] Hartshorne observes that 'aesthetic value has an intrinsic and an instrumental value . . . [where] subjective does not mean relative or a mere matter of opinion' (1997: 208). Following Whitehead, the 'actual satisfaction' of an occasion just is a 'positive value in reality' (Hartshorne 1997: 208). The challenge is in understanding its instrumental value. Indeed, counter-intuitively, 'it is objective value that may indeed be partly a matter of opinion' (1997: 208).

Any effort to base an ethics on this experience of beauty invariably runs up against the challenge of translating from the intrinsic to the normative: whose beauty? Beauty how? Beauty when? The eye of the beholder haunts any ethics from beauty if we presume that an ethics must provide a normative heuristic for judgement of discrete actions or characters. This is not to say that such attempt is misguided, but rather that it must be accompanied by a correlative rethinking of what ethical theory aspires to do. It must be an ethics in the immanent, Spinozist sense, not the juridical Kantian sense.[30]

Efforts to extract a representative axiology also risk insufficiently considering the differentiation between the intensive and the extensive. For both Whitehead and Deleuze, it is *how* an occasion becomes and *how* individuation proceeds that determine beautify-ing, intensify-ing, harmonis-ing or discord-ing qualities and effects. Beautifying is relational and will vary contextually and situationally. When these criteria (beauty, intensity, harmony) are presented as stable entities in the extensive realm, it contributes to a sense that philosophical thought can pursue judgement of entities *independent from their genetic processes*. It is a question of a choice in emphasis. Do we approach value-ing from a primarily conceptual lens with the desire of articulating a stable heuristic for judgement (as in Hartshorne's 'Neo-Classical Theism') or do we approach in terms of thinking conceptual effects on experience (emphasising the radical empiricist side)? From this latter emphasis, how does thinking a subject as a 'society' of occasions entangled along different lines (the rigid and molar, the supple and molecular, and the creatively destabilising line of flight) alter orientation towards prevailing normativities?

Asking this question challenges presumptions of subjectivity as reflective of a stable inner essence. This presumption reinforces a tendency to conflate what is possible or real with what is presently deemed 'normal'. While this conflation undoubtedly has psychological and sociological aspects, it is also a consequence of how metaphysics partially structures and influences habits and activities of attention. One manifestation involves how the production of norms based on statistical samples is part of a feedback loop that perpetuates itself into the future through guiding attention.

I am not claiming that statistical analysis is always or necessarily productive of future conformity. It is a question of *how* procedures of statistical science are used. Though statistical analysis and claims should most rigorously be understood as exclusively descriptive, this fails to reckon with how they interact with the processes of attention through which a metaphysical 'society' is constituted. The subject, as historical route of occasions, can in part be described as a *manner of attention*. But each occasion of attention is not fully determined historically, and therefore has the potential to either expand or diversify the repertoire of this manner or to solidify or attenuate its capacities. How we use the powerful information gathered through statistical analysis can impact such attention in important ways correlative with the tension between conformity and novelty that all enduring subjects negotiate. While a certain degree of 'conformity' is constitutive of any stable society-subject, when such conformity enforces imposition couched in appeal to consensus or normative equality, it is likely to hinder rather than inspire creative becoming.

This is why Deleuze is sceptical of 'consensus' as philosophical norm. When philosophical thought is oriented towards a normative assumption of communication, it 'only works under the sway of opinions in order to create 'consensus' and not concepts. The idea of a Western democratic conversation between friends has never produced a single concept' (WIP: 6).[31] The presumed ideal of communication is easily enlisted by procedures of power to dismiss outliers or reinforce the status quo. Similarly, when norms are bolstered by the authority of statistics in an uncareful way, this conflation can be leveraged by power structures – whether socially, politically or subjectively – to deny the need for attention to the unusual in the name of 'how it is'.

A process metaphysical view of subjectivity intensifies the stakes of such loops because the quality of future occasions is directly impacted by their manner of prehending the past and present. Attention that dismisses the unusual because of a conflation between expected patterns and a presumed stability of how it is can serve to diminish the range of contrasts ingredient to an occasion. This has the effect of impoverishing the society's capacity for creative or novel response because its range of experience has been diminished. A processual view of subjectivity seriously problematises complacent appeals to 'how it is' when these are underwritten by a closed loop rejection of 'datum' that does not fit the claim in question.

The implications go in several directions. For one, a processual view implies a different orientation towards the very question of normality. 'Is this normal?' dissolves as a pseudo-question if one asks it to find a bolster of security that the future *will necessarily follow as expected according to present norms*. The import of the question changes, such that there is a way in which the answer is always 'yes' and sometimes 'no'. Yes, if we are asking about any event or occasion, it is 'normal' in the sense that, because it has happened, it is an expression of real potentials in a complex convergence of real conditions: affective, material, conceptual, ideal, imaginary, and so on.[32] This convergence results in an expression that does not stand still, that is gone, carrying on, taken up by the next achievement of 'normalcy'.

This significantly ironises the meaning of the word normal in a manner reminiscent of Dan Smith's remarks about the two poles of Deleuze's philosophy: '"Everything is ordinary" and "everything is singular"' (Smith 2012: 115). With regard to attention, the motivation of the question of the normal is crucial. Nothing is 'normal' if the term is understood as comprehensive justification, excuse or valorisation. In this sense, the danger of the ascription of the normal is its enlistment as mode of dismissal, *a gesture towards the non-necessity of paying attention*. A day just like any other day is a day where the subject meets the unknown.

As forms of abstraction, perceived norms (whether about the normal, expected or even the good) can reinforce a static sense of identity when they *discourage rather than encourage* attention to the particularity or singularity of an encounter, event or occasion. This does not of course mean that they can just be thrown out, and they can also be used to encourage or intensify attention. Here we might think of the intensity of ritual forms in some religious practices as a matter of intensifying attention rather than diminishing it. In this sense, discipline need not always be oriented towards mastery or prediction, it can also be a form of micro-discipline towards attention to the particular. Whitehead reminds us that there are, in principle, no limitations to the rewards of such attention: 'we are instinctively willing to believe that by due attention, more can be found in nature than that which is observed at first sight' (CN: 20). By the time of *Process and Reality*, this instinctive belief in the potentials of attention is coupled with a metaphysical claim: 'the complexity of nature is inexhaustible' (PR: 107).

The stakes of attention, of whether or not one's orientations towards perceived normalcies intensify or diminish attention, are bound up with potential for complexity and transformation. Whitehead does not hesitate to link evolutionary transformation to attention: 'Evolution in the complexity of life means an increase in the types of objects directly sensed' where 'objects' are abstract entities inclusive of, for example, musical themes, conceptual constellations, images, and so on (CN: 104). Attention develops through its abilities to make finer and more subtle determinations, a capacity that is strengthened through an orientation that deems *such more fine-grained determination always possible and never finished*. In this way, the lived reality of a perceptual universe expands (or does not): 'Delicacy of sense-apprehension means perceptions of objects as distinct entities which are merely subtle ideas to cruder sensibilities. The phrasing of music is a mere abstract subtlety to the unmusical; it is a direct sense-apprehension to the initiated' (CN: 104).

Whitehead's example risks being reified if taken too literally, especially since it operates with an implicit reduction between the unmusical and the initiated (as if there were one initiation, one music and one species of the unmusical). Nevertheless, his point remains operative, provided we expand the potentials of delicacy to resist straight-forward linear hierarchy: the intensities of such more delicately attuned sense-apprehension can emerge in differently modulated contexts, species, cultures, and so on. This also applies to the development of norms as abstractions guiding behaviour. Rather than universal concepts that cover all particular actions without any change in their own nature, within a process view, any general concept or norm is liable to transformation insofar as events exceed or reorganise the boundaries of the category in question. Each event or occasion is potentially different and may cause a reorganising or redefinition of whatever collective category is used retroactively to describe it.

This does not mean that all persistence or general patterns are eradicated in a ceaseless and chaotic flow. The issue is one of tendencies and perspectival scales. Permanence, averages, expectations are by degrees. Such a view thus challenges insistence on *metaphysically* determinate identities advanced in the name of objectivity or order. This insistence often takes sociological or psychological forms when it presumes that any giving up of determinate identity can only result in a chaos of instability – a classic deployment of an all-or-nothing logic. Massumi describes this, following Jean Oury, as 'neurotic normativity, which invests itself body and soul in the compulsion to repeat the same' (2014: 70).[33] Such 'normopathy' is incapable of living in the paradoxes that a process metaphysical conception of subjectivity enacts. Opening to the creative potential of each actual occasion is existentially impossible without a willingness to tolerate ambiguity, since the occasion operates on the cusp of sedimented or stable values.

Each actual occasion, in its concrescence, contains a mixture of different potential tendencies. In this sense, if the subject becomes more attuned to micromovements and segmentations-in-progress of the second line, this will be correlative with an experiencing of normative ambiguity from the perspective of the molar. Creative possibilities of the occasion are invariably bound up with destructive possibilities as well. It is a matter of intensity, dosage and degree, not categorical separation. The art of attention thus involves learning how to respond so as to accentuate or develop those tendencies best situated to achieve the occasion in a creative fashion. This 'logic of mutual inclusion' heightens the stakes of attention since acts of attention are partially constitutive rather than merely receptive. Moreover, attention itself is no longer a univocal term, since it is not just what we 'pay' attention to, but *how* we attend, the style and manner of attending, that is part of the relational constitution. This challenges the zealous defender of a reality made up in advance who seeks to insulate themselves against these stakes and risks. It challenges the 'normopathy' that insists on a clear and determinate divide between the 'creative' as play and the seriously descriptive as work. Such an orientation:

> magnifies the minimal difference opened by the paradox of play into a monumental difference that is taken overseriously. The gap is erected into a structural divide, which is defended at all costs in the name of 'the way things are.' No mixing allowed: fight or play, but for sanity's sake don't contrive to do both at once. (Massumi 2014: 70)

Returning to statistical or probabilistic averages, the question is how these are used to direct attention. While statistical data are useful both descriptively and predictively, these retroactive reports easily become self-fulfilling prophecies when coupled with the prevailing acceptance of essentialist substance metaphysics. When the data in question are taken to characterise predicates of a stable formal identity, they have the result of directing attention into the future by selecting what is expected and discounting, or, more likely, not even noticing, that which does not fit the expected pattern.

This is particularly important to think about as algorithmic data-mining becomes such a pervasive feature of digital ecology. Such predictive technologies, whether in interests of surveillance, security or profit, use accumulated actions of the past to predict the likely actions of the future.[34] Franklin Foer observes: 'the whole effort [of automation] is to make human beings predictable – to anticipate their behavior, which makes them easier to manipulate' (2017: 77). Efforts at predictability are correlative with a massive digitalisation of information where the new limited resource becomes *attention*. Foer reports that between 2006 and 2012 'the world's information output grew tenfold' (2017: 88). This growth of (a certain style of) information challenges attention: '[information] consumes the attention of its recipients . . . a wealth of information creates a poverty of attention' (cited in Foer 2017: 88).

Calling 'data the new oil', Foer emphasises how this dynamic becomes a tool of control for dominant corporations in a capitalist context. Mark B. N. Hansen develops a more metaphysical analysis, drawing specifically on Whitehead. Hansen understands that from a processual perspective, acts of attention in the present 'feed-forward' (his language) into the future. The future is created through activities of attention in the present. However, he correctly seeks to de-phenomenologise this claim. This is not a claim purely about so-called 'subjective' (in a human sense) attention, nor is it a psychological claim. It is metaphysical. The same structure which creates the danger that Foer and others are alarmed about –a danger that becomes all the more pronounced in the context of polarised information and media landscapes where like-minded views cluster together in echo chambers of confirmation bias – also contains the possibility for what Hansen calls its 'recompense' (2015a: 24, 51–3).[35] Indeed, it is because 'there is always more sensory potential to data than what gets captured by the techniques central to today's culture industries' (Hansen 2015a: 66) that Hansen argues for the possibility of an alternative 'feed-forward'.

The existential locus of such an alternative is habits of attention. If we are to challenge closed loops of predictive analysis and the 'narrow instrumentality of capitalist cultural industries' (Hansen 2015a: 66), habits of attention must attend to that which escapes the norms, escapes probabilistic prediction and expresses novel or unforeseen potentials inherent in the present. This is what Deleuze and Guattari have in mind when they proclaim 'the importance of statistics, *providing it concerns itself with the cutting edges and not only with the "stationary" zone of representations'* (ATP: 219, emphasis added). These cutting edges include rather than omit anomalies and potential points of transition between molar forms into something else. The powers of statistics are not best construed as representing 'how things are', but as tools for developing a sense of what connections are possible. The goal is not representation but construction.

A processual view understands that statistical averages do not represent that which is as a reality made up in advance, but rather express the convergence of particular patterns of occasions through the prism of a selected perspective and according to a particular aim or goal. Such averages, especially when they get converted into norms, do not explain or justify, but instead must be approached in an attitude of caution and attention. To paraphrase Deleuze, averages do not explain, but are what must be explained.[36] What acts of attention and guiding aims produced the averages in question? How does a different activity of attention challenge the perceived stability of such averages? What cutting edges can be extracted from them, not so as to perpetuate their depiction of how things are but so as to utilise or leverage their pattern into a different becoming? This does not mean we dismiss the extent to which sedimented ascriptions of 'the normal' inform present behaviour in often unconscious ways, nor deny that categories derived from statistical averages enact real material effects. But it is to raise awareness about the way that analyses that are not careful to attend to the metaphysics framing their enquiry risk reifying into that which they seek to undo: structures of identity that block rather than accommodate or encourage a greater expression of potentials.

Notes

1. Though pronounced as dismissal, Rorty's remarks in a 1983 *Times Literary Supplement* review express this metaphysical convergence coupled with stylistic divergence: 'Deleuze's account of "desiring machines" [in *Anti-Oedipus*] is like the account of "actual entities" in Whitehead's *Process and Reality*, rewritten by somebody whose favorite poet is Brecht (rather than, as Whitehead's was, Wordsworth)' (Rorty 1983: 619–20).
2. It does of course function as a relatively stable metric relative to this 'cosmic epoch' as Whitehead puts it, which is 'formed by an "electromagnetic" society, which is a more special society contained within the geometric society' (PR: 98). This leads into Whitehead's most technical thinking in developing an abstract account of extensive measurement that does not presuppose *as necessary* Euclidean, or indeed any particular, geometry.
3. Another less intuitive but relevant example might be one's visitation and participation in the 'individuation' process that is a (relatively) enduring website.
4. Stengers of course is not the only figure to develop the *pharmakon* in innovative ways. Most widely associated with Derrida's reading of the *Phaedrus* in 'Plato's Pharmacy', it is also used by Bernard Stiegler. Derrida and Stiegler are primarily interested in thinking about how technologies (writing for Derrida, contemporary media for Stiegler) function as both poison and then remedy for the poison they have introduced (writing takes away memory, but then gives it back to us in a different form). Stengers is more interested in developing its consequences for *attention*. See Derrida 1981; Hansen 2015a: 50–5; Stiegler 2014.
5. For Whitehead, this second is the organising or cohering of certain actual occasions as 'societies', discussed in Part II below. For Deleuze, this second involves the intensive processes that give rise to extensive individuals.

6. See Krauss 1998: 21–6 for helpful discussion of this point. The 'beads on a string' model, criticised by Bergson as a spatialisation of temporality, is very difficult to avoid and structures a way of thinking actual occasions as little units or atoms. This is prevalent in the literature; indeed some commentators, most notably Kline, ascribe it to Whitehead himself, accusing him of a 'cryptosubstantialism'. I am inclined however to agree with Judith Jones that this reading is a result of how deeply embedded metaphysical habits are in the grammar of the language Whitehead has at his disposal. See Jones 1998: 85–8; Kline 1983.

7. As per Hartshorne's 'zero fallacy', this tendency towards minimal creation should not be assimilated into a total negative claim.

8. I agree with Hansen, Debaise, Jones and Stengers that it is a mistake to conflate experiential or phenomenological events with metaphysical actual occasions, but I further agree with Jones that the phenomenological must be coherent as a species of the metaphysical (1998). Though there is no evidence of Whitehead being influenced by the phenomenological tradition per se, Jones notes the 'unfortunate' lack of such engagement (1998: 49).

9. To my mind, this leads to Whitehead's worst book displaying a Euro-centric notion of the spread of civilisation that is dubious at best. However, Whitehead's shortcomings in this regard also illustrate the extent to which his speculative cosmology exceeds his own empirical vantage. Indeed, in one sense, the rise of decolonial thinking, at its best, exemplifies the adventure of ideas.

10. I refer to Whitehead's four stated criteria for a speculative philosophical scheme: applicability, adequacy, logic and coherence. The first two express the 'empirical side' of a metaphysics and the second two the 'rational side' (PR: 3). Applicability refers to the manner in which the abstract concepts of a system are meaningfully applied to interpretation of concrete experience and reality. Adequacy complements this applicability to say there are no known items of experience or reality incapable of interpretation through the abstract framework.

11. Detailing this typology, Whitehead stresses that he is 'deserting metaphysical generality' and instead basing his characterisation on 'our present epoch' (PR: 96), that is a description phrased according to empirical concepts drawn from the current epoch's conceptual map. In calling attention to this distinction, Whitehead implies the possibility of alternative epochs while likewise characterising 'our present cosmic epoch' as 'dominated by a society of electromagnetic occasions' (PR: 98). He thus privileges the physical sciences as foundational for characterising the assumptions that dominate *this* epoch while simultaneously alerting us to the inherent fallibilism of any such privileged scheme. This implies alternative epochs as real potentials for the future.

12. I use the language of 'set' loosely. I am allergic to set theory as a formal language for thinking Whitehead or Deleuze because it is defined by axiomatic definitions that encourage a thinking of sets as essentially fixed and a-temporally determinate. Dan W. Smith's work contrasting an 'axiomatic' philosophy of mathematics (Badiou) to a 'problematic' (Deleuze) remains the crucial intervention on this issue (Smith 2012: 287–311).

13. This could be contrasted with a nexus of occasions constituting the global reality of a given day chosen from the Gregorian calendar: September 11, 1974. This is a nexus, but there is no essential connection in terms of a mutually reinforcing and sustaining *quality* for this nexus other than the one chosen by the calendar. It does not acquire a social order and is not a society.

14. Whitehead clarifies that a society with a personal order need not have consciousness (PR: 35).

15. Whitehead also offers further distinctions between *structured societies, subordinate societies* and *subordinate nexus* (PR: 99–100). These distinctions are not relative to scale, but rather describe their respective relational dependence. For example, a subordinate society exists in a relation of dependence within a larger structured society. Though it informs the structure of that society it is capable of existence or survival independently

of it. Whitehead's example is a molecule within a cell – the molecule has an integrity of its own, even as it is also indisputably implicated in the functioning of the cell. This metaphysical relation (it need not be physical, this is only an example) is scale flexible. The individual human is a structured society, but it is also a subordinate society in relation to larger socio-cultural processes termed 'society' in the conventional sense. By contrast, what Whitehead calls a 'subordinate nexus' *could not exist* without the structured society in which they occur. Whitehead gives the 'empty space' within a cell as an example of a subordinate nexus, since if the cell were to disintegrate this space would also collapse. A river is a structured society, while the water flowing in it, as such, is a subordinate nexus.

16. Whitehead's employment of a rhetoric of 'higher' and 'lower' stands in performative tension with his enlistment in 'posthuman' discourses. While this bears additional scrutiny, a few brief caveats may distance Whitehead's language from the traditional Chain of Being. For one, the distinction between higher and lower societies is not necessarily equivalent to conventional species distinctions or standard biological taxonomies. Additionally, Whitehead's higher and lower is meant only to refer to the potential for complexity of experience, not to endorse any moral hierarchy.

17. 'Supernormality is an attractor that draws behavior in its direction . . . [and in this sense] it is a force not of impulsion or compulsion, but of affective propulsion' (Massumi 2015: 9). That is, it involves the activation of an affective lure that can be satisfied in differing ways.

18. Malabou considers the political dimensions of how uncareful valorisation of flexibility is easily enlisted into capitalist multitasking and short-term labour. The question then is how 'plasticity' is used. She worries that insufficient attention allows 'neuronal plasticity . . . [to] justify a certain type of political and social organization' (2008: 9).

19. In an instance of a metaphor with implications beyond the author's intention, Whitehead observes that 'A vegetable is a democracy; an animal is dominated by one, or more centres of experience' (MT: 24).

20. The relation between the organic and inorganic is also not exclusive, as Whitehead notes: 'We do not know of any living society devoid of its subservient apparatus of inorganic societies' (PR: 103).

21. Massumi points out that *fuite* 'covers not only the act of fleeing or eluding but also flowing, leaking, and disappearing into the distance' (ATP: xvi). He also stresses that it has no relation to flying.

22. DeLanda characterises 'plane of consistency' as largely synonymous with 'intensive spatium' (DR), 'plane of immanence' (WIP) and 'Body without Organs' (ATP, AO). The plane of consistency has a thermodynamic sense in referring to 'intensive properties like pressure, temperature, or density' (2002: 199). It also has a sense derived from 'properties of ordinal series' in which what is important is asymmetrical relations between series – differences between differences if you will. A shift in one series reverberates through others and produces alterations, changes, differentiations. Finally, DeLanda states that 'difference, distance, and inequality' are the positive characteristics of the 'plane of consistency' (2002: 199). DeLanda's account emphasises the material dimension of the plane of consistency as the condition for the plane of organisation. The question arises to what extent a line of flight is something other than a shift in perspective. If plane of consistency is construed in material terms, then it is, in a sense, always already present and active, and it is a question of finding a line of flight to bring us into contact with it.

23. This indicates the complex entanglement between descriptive and prescriptive in the schizo-analysis texts (an entanglement present in Deleuze's earlier works also). There cannot be a sharp line between a 'neutral' 'representation' of the real – as in the dogmatic image of thought's understanding of metaphysics as *a priori* armchair exercise – and effects of thinking as part of the expressive processes of the real.

24. This is a region of resonance with Foucault, whose work, especially *Discipline and Punish*, offers an analysis of how power develops strategies of control that operate at the level of the molecular.

25. Whitehead clearly proposes such teleology: 'The teleology of the universe is directed to the production of beauty' (AI: 265) and Henning is right that any viable ethics must be consistent with it. The question has to do with how, or even if, we might consistently represent such an ethics within the assumptions of traditional ethical theory.

26. Though they title a plateau: 'How do you make yourself a Body without Organs?' they stress the care and caution with which this should be approached. On the one hand, 'it is easy' and 'we do it every day', on the other hand, it requires 'caution' (ATP: 159–60).

27. Whitehead's own problematic espousal of 'progress' serves a cautionary note here. He praises the nineteenth century as 'an epoch of civilized advance – humanitarian, scientific, industrial, literary, political' (AI: 278). This praise is easily complicated by juxtaposing 'humanitarian' advance with the genocide of the North and South American Indigenous populations, 'scientific' advance with nineteenth-century discourses of eugenics and racial hierarchy, and 'industrial' advance with the proliferation of child labour and Gilded Age robber barons, not to mention widespread environmental exploitation and pollution.

28. For more on the 'beautiful soul' as a possible divergence between Whitehead and Deleuze, see Duvernoy 2019.

29. The diagram is the creation of Hartshorne, German writer on aesthetics Max Dessoir, and Kay Davis, an artist and student of Hartshorne's at Emory. Of note is the placing of 'Beautiful' at a point of tension between 'too much' order and 'too little', which Hartshorne attributes to Davis. Prior to her insight, beauty was placed at the top of the vertical and aligned with maximum order, reflecting the rationalist inclinations of Hartshorne. However, with Davis's help, he comes to realise that 'unless there is some aspect of freedom, disorder, conflict, uncertainty, unpredictability, there is no intensity of experience' (1997: 204).

DIAGRAM OF AESTHETIC VALUES
Undiversified unity, absolute order

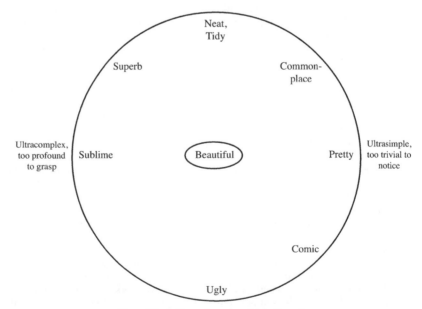

Ununified diversity, absolute disorder

30. Dan Smith's essay on Deleuze's 'immanent ethics' is the definitive commentary on this issue (Smith 2012: 146–59). Following Spinoza and Nietzsche, and in contrast to rules-based moralism, an immanent ethics 'thinks according to the immanent mode of existence it implies' where mode of existence is an ontological notion (2012: 147).

31. Deleuze's rhetoric here is liable to be overstated, especially if one does not realise its context of criticising Rorty's linguistic conception of philosophy.

32. Auxier and Herstein provocatively characterise this point: 'What is actual is possible. This is the first law of metaphysics' (2017: 5).

33. Founder of the experimental La Borde psychiatric clinic where Guattari worked, Jean Oury is an important figure in the French intersection of phenomenology and psychiatry and author of several important texts not translated into English, most notably *L'Aliénation* (Paris: Galilée, 1982) and *Création et schizophrénie* (Paris: Galilée, 1989). For an introduction, see Oury et al. 2007, 'The Hospital is Ill', *Radical Philosophy*, 143.

34. See Hansen 2015a and 2015b for discussion of Whitehead and data-mining technology. See Scranton 2015: 83–8 for discussion of self-fulfilling prophecies informed by such technology.

35. Such recompense involves humans taking the sensibility produced by digital technologies as means for 'enhancing the intensity of experience' (Hansen 2015a: 24).

36. Deleuze comments 'I have always felt that I am an empiricist, that is, a pluralist . . . this equivalence between empiricism and pluralism . . . derives from . . . [a] characteristic[s] by which Whitehead defined empiricism: the abstract does not explain, but must itself be explained' (Deleuze 1987: vii); See also his remark: 'The first principle of philosophy is that Universals explain nothing but must themselves be explained' (WIP: 7).

Part II
Applied Metaphysics and Existential Implications

Part II

Chapter 5

Attention, Equivalence and Existential Territories

Existential territories and psychic ecologies

As is well known, ecology and economy share a root (οἶκος) referring to the home, family or family's property. This root expresses what becomes a persistent entanglement between ecological theorising and social ideology – from Social Darwinism to neoliberal selfish genes.[1] Given the dangers of such conflations, it is tempting to insist they represent only errors, expressions of pseudo-science enabled by a combination of epistemic weakness and opportunistic greed. This compartmentalisation is too simple. While we can certainly bemoan the failures of particular conflations as well as distinguish between levels of nuance in a theoretical position and its reduced or trivialised social expressions (which nonetheless often have real effects in behaviours and choices), these conflations point to questions inherent in the concept of ecology itself. Scientific ideals of dispassionate neutrality notwithstanding, ecological theories must entail, at some level, implications for human life. It is true that these implications rarely translate in manners as simple as popular ideologies would have them, but this alone is not enough to foreclose the question. *How* we understand the order or *logos* of the οἶκος necessarily has normative implications.[2] Though these implications may not be definitive, the entanglement is part of the stakes of ecological thought as such.

Recent decades have seen proliferation in the deployment of 'ecology' as a term varying in precision and technicity. Following Deleuze, one source of confusion arising from the proliferation of ecological discourses is lack of distinction between philosophical ecology and scientific ecology.

To develop this distinction, it is useful to briefly review the term's modern origin.

The first modern usage of ecology was by Ernst Haeckel in 1866, who defined it as 'the whole science of the relations of the organism to the environment, including, in the broad sense, all the "conditions of existence." These are partly organic, partly inorganic in nature' (Stauffer 1957: 140). The date of Haeckel's proposal is no accident, coming closely after Darwin's publication of *On the Origin of Species* (1859).[3] Indeed, Haeckel's emphasis on *relations* is inextricable from evolution as *a new theory of change*. The coupling is intriguing: Darwin's theory of evolution drives interest in thinking about external factors and variables shaping evolutionary fitness. Evolutionary biologist Leigh Van Valen for example writes, 'evolution is the control of development *by ecology*' (1973: 488, emphasis added). Theories of evolution are also one of the impetuses for the interest in metaphysics of process that emerge in the later nineteenth and early twentieth centuries.[4] These two themes (change and relations) thus already point to intersections between the science of ecology and process metaphysics. Even while the relevant elements (populations, organisms, genetic material, chemical elements, or other biotic or abiotic factors) vary, ecology is concerned with understanding, mapping or measuring how different components or variables effect and relate to one another. Such relation involves complex loops of causality and dynamic processes.

An intersection or resonance is not of course an equivalence. So, while Haeckel's definition asserts that ecology considers 'all' conditions of existence, scientific practice invariably limits inclusion to conditions deemed suitably relevant. This limitation is most neutrally described as practical following operative procedures of science. Scientific emphasis on precision, unambiguous and preferably quantifiable observation data, and replication of results entails that ecological sciences have to delimit relevant variables in some way. Such restriction is then constitutive of a specialisation, whether in terms of 'levels' (organism, population, community, ecosystem) or 'domains' (spatial ecology, landscape ecology, physiological ecology, evolutionary ecology, functional ecology, behavioural ecology, and so on).

While this restriction of relevance to enable testability applies to all empirical sciences, ecological sciences present particular difficulties. These follow a constitutive methodological tension: the limitation of variables *in practice is balanced by an aspiration towards maximal inclusion in principle*.[5] As science, ecology is defined by this desire towards maximal inclusion of all *relevant* conditions, but such maximal inclusion as an ideal raises methodological and conceptual challenges. Of particular note are the 'complexity problem' and the 'uniqueness problem' (Sarkar 2005).[6] Ecological systems are complex by nature and it appears likely that they exhibit emergent qualities at different scales and are, in other words, more than just the sum of their parts. Moreover, and more pertinently for my concerns here, because of this complexity ecological systems are to some extent unique.[7] Sarkar presents these as problems for the scientific practice of ecological sciences, but they also show the extent to which the limitation of relevant conditions is entangled with metaphysical presuppositions that remain largely structured around the 'bifurcation of nature'.

Given the entrenchment of this bifurcation, the speculative proposals under development here remain extraneous to scientific ecology as currently practised. This makes it all the more imperative to insist on the difference between science and philosophy, a difference that Deleuze frequently explores. For Deleuze, this difference is best summarised in two ways: science is oriented towards the creation of functions as 'propositions in discursive systems', whereas philosophy *creates concepts* (WIP: 117). The practice of science is oriented towards establishing functions that provide a plane of reference linked to predictable states of affair. It freezes continuous flows and instead presents variable states of affairs as discrete or static pictures: 'science is like a freeze-frame . . . It is a fantastic slowing down . . . a function is a Slow-motion' (WIP: 118).[8] The slowing down expresses how scientific propositions isolate particular variable relations as a means of fixing references to be studied.

In this regard science represents an important and powerful approach to understanding reality, but not reality itself. When Deleuze says that reality has an 'infinite speed of birth and disappearance', this is from the perspective of metaphysical inclusion encapsulating all levels and scales – from the sub-atomic to the cosmic (WIP: 118). From such a vantage, solidities or appearances are born and disappear, at different scales, constantly. Deleuze also uses 'chaos' to refer to this perspective prior to the establishment of a fixed plane of reference or consistent ordering of events, describing it in metaphysical language: 'it is a void that is not a nothingness . . . containing all possible particles and drawing out all possible forms, which spring up only to disappear immediately, without consistency or reference' (WIP: 118). Both science and philosophy engage this virtual chaos, but they do so with different aims and strategies.

In contrast to science's creation of functions to slow down the chaos by producing fixed planes of reference, a philosophical creation of concepts *gives a consistency* to the chaos through conceptualising *events*. Whereas functions require independent variables as elements to establish discrete relations, concepts are defined by an 'inseparability of variations' (WIP: 126). The concept therefore is oriented towards continuity, but it is a continuity that expresses itself variously. To use a spatial model, consider the morphology of a sphere as it is twisted, pulled, compressed, elongated, and so on. We might consider paradigmatic philosophical concepts (the Good, the Real, the True, Beauty, Justice) as such constructions that give a coherence or consistency to the chaos of the real, but a coherence that changes forms and manifestations with changing conditions. In this sense, rather than a 'freeze-frame', the concept is 'formed like a consistent particle that goes as fast as thought' (WIP: 118). This amounts to a different approach:

> through concepts, philosophy continually extracts a consistent event from the state of affairs – a smile without the cat, as it were – whereas through functions, science continually actualizes the event in a state of affairs, thing, or body that can be referred to. (WIP: 126)

Of note is Deleuze's alignment of the concept with events – rather than a thing, the concept is a manner of achieving consistency, a kind of event. It selects from this chaos. 'The Good' for example, is not a static form, but a style of event ('a smile without the cat'). Moreover, concepts have variations, components, internal consistencies that shift over time, that become differently in passing thresholds to produce new variations (WIP: chapter 1).

What are the variations in *the concept* of ecology? We have seen that its components involve an inseparability of the following variations: (1) how we construe the whole, i.e., what *counts* or *matters* for inclusion, (2) how we characterise the specific or singular 'elements', (3) the relations between, and (4) how these relations move or change over time. If a scientific practice of ecology agrees with these components in principle, a philosophical conception of ecology that resists presumptions of the bifurcation of nature cannot *begin* with an enforced distinction between the geophysical and the social or psychic. It must endeavour to respond to Haeckel's definition regarding *all the conditions of existence* without presuming this only applies to the physical or the 'primary qualities' of the physical. This does not dismiss the need for perspicuous attention to material interactions. It is not an idealist point. The idea is that a philosophical concept of ecology maintains the commitment to maximal inclusion (a commitment already displayed in James's radical empiricism) while bracketing metaphysical presuppositions. Material effects, properties and actants are certainly to be included, but so are extra-material concepts, ideological infrastructures, social systems, and psychic and affective habits and patterns.

In this regard, Guattari's 1989 *The Three Ecologies* proposes three 'ecological registers' (material, social and psychic) that follow a dynamic logic of interrelation, represented here by overlapping circles in a Venn diagram:[9]

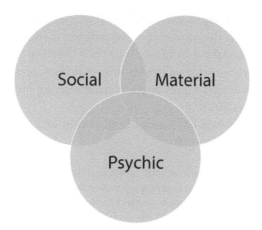

Experience emerges out of the overlap between all three that Guattari calls an 'existential territory' (2008: 23). The metaphysical status of this overlap and territory is not exclusively psychological. The neologism here is purposeful in endeavouring to rethink this overlap prior to the assumption of a discrete human subject that lives it.

The relations (an operative component of any concept of ecology) between these registers are reciprocal and horizontal rather than hierarchical: material environments inform social relations inform psychic habits and desires that act on material environments, and so on. Psychic habits emerge out of social relations as well as partially influence and shape them. None function as independent systems; rather, they are constituted through ongoing processes that produce divergent qualitative expressions. Guattari is thus working against tendencies of first philosophy and architectonics oriented by first principles or primary foundations. In this sense, while Deleuze and Guattari draw deeply on Marx's analyses of capitalism, they reject *the material* as the sole exclusive relation that explains all other relations.[10] This is not to deny its importance. And it is also to reject the psychic as an opposing exclusive relation that explains all other relations. This is part of the concept of ecology wielded here: there is no single relation or domain that explains all others. And yet, there is also no relation or domain that is merely extraneous or negligible.

Guattari's work, notwithstanding the notorious idiosyncrasies of his terminology, anticipates central insights in recent work in political ecology (Jane Bennett) and the rising prominence of 'integral ecology' following Pope Francis's second encyclical *Laudato Si': On Care for Our Common Home* (2015).[11] Francis's fourth chapter *Integral Ecology* includes numerous claims resonant with Guattari. This is likely to surprise those who remain attached to ideological identities: Guattari, the fiery activist and critic of all forms of 'repression', and the Pope, institutional head of what has sometimes been the most repressive political institution in the world, making consistent claims? In addition to their differing contexts, this is an example of the way in which alliances more readily emerge when ideological identity is no longer a presumed or fixed starting point (see Braidotti 2013 and Stengers 2015).

For example, Francis links social dysfunction and environmental devastation as closely connected:

> We are faced not with two separate crises, one environmental and the other social, but rather with one complex crisis which is both social and environmental. Strategies for a solution demand an integrated approach to combating poverty, restoring dignity to the excluded, at the same time protecting nature. (2015: 139)[12]

Like Guattari, the Pope asserts fundamental interconnection between 'ecology' as 'studying the relationship between living organisms and the environment in which they develop' and the necessity for 'reflection and debate about the conditions required for life and survival of society' with regard to 'certain models of development, production and consumption' (2015: 102). In all of these cases, an ecology does not presume a fully determinate distinction between nature and culture, between a 'natural' environment and a constructed one. Francis observes, 'Nature cannot be regarded as something separate from ourselves or as a mere setting in which we live. We are a part of nature, included in it and thus in constant interaction with it' (2015: 104). Similarly, Guattari insists that 'now more than ever, nature cannot be separated from culture' (2008: 29).[13]

Guattari presciently alludes to the extent to which the collusion of techno-science, biotechnology and Integrated World Capitalism (IWC) – not to mention destabilisation induced by the intersection of climate change disaster, resource depletion, and displaced peoples as a result of military or climate crisis (all trends that have only intensified since the time of Guattari's text) – make any sharp distinction between the natural and the cultural a relic.[14] Genetic modification, cloning, projects in insect cyborgs,[15] trace amounts of radiation that have covered the planet since the nuclear age, or the ongoing influence of greenhouse gases on planetary weather systems all demonstrate mutual implication. Conversely, 'nature' is a consequential player in human events – think of rising numbers of displaced peoples as a result of climate crises or the potato blight in mid-nineteenth-century Ireland that led to waves of emigration, or more recently the correlation between unprecedented drought in the Syrian countryside and the ensuing civil war and ongoing refugee crisis.[16]

This mutual implication is demonstrated in the much-debated naming of the Anthropocene as a new geological era that acknowledges human industrial activity as an unprecedented influence on global climate.[17] However, at least as early as Darwin there was recognition that an exclusive distinction between human 'culture' and nonhuman nature was strained at best. Darwin's 1881 study of earthworms reflects that worms, through their physical construction of viable fertile topsoils, are material enablers of sophisticated agriculture and therefore, 'Worms have played a more important part in the history of the world than most persons would at first assume' (quoted in Bennett 2010: 95). Not only do worms make possible 'seedlings of all kinds', but they also help preserve past artefacts of history 'for an indefinitely long period every object, not liable to decay, which is dropped on the surface of the land, by burying it beneath their castings' (2010: 96). As Jane Bennett summarises, 'worms participate in heterogeneous assemblages in which agency has no single locus, no mastermind, but is distributed across a swarm of various and variegated vibrant materialities' (2010: 96). In such assemblages, the line between 'cultural' and 'natural' agents and activities is not easily drawn. Even if we insist on some version of this distinction, the pace and scale of their co-implication are implicated in a loop. We may be able to retroactively make distinctions on a continuum with culture on one side and nature on the other, and these may serve certain pragmatic purposes in some contexts, but the increasing pace of their co-implication problematises any hard and fast *metaphysical* distinction.

This point has been made from a variety of disciplines and orientations in recent years.[18] Because of its fashionability, it is important to stress its limits. Challenging the metaphysical distinction between culture and nature is emphatically not to deny that there is a more-than-human material world that informs and constrains human possibilities. Though humans participate in nature, the materiality of the real exceeds human control *to some degree*. To recognise human and nonhuman entanglement is not to deny the possibility of nuanced qualitative distinctions. Pushing on the culture/nature distinction does not deny that there are better and worse forms of material practice for the future of life, its qualitative expressions, diversity, complexity and richness. The idea is rather that insisting on a stark demarcation between culture and nature either encourages a lack of consideration of the relational consequences of such material practices, or, alternatively, reinforces a sense of fatalism, often theologically or religiously inflected.[19]

This is not a claim to the 'end of nature' as such – rather, it is critique of *a concept* of nature that reinforces false separations. In this regard, it is interesting to note awareness in comparative philosophy that many Indigenous languages and traditions (often romantically caricatured as the 'most in touch with nature') do not conceptually mark this separation (between nature and culture). Thomas Norton-Smith for example notes that 'there is no analogue of the concept of the natural world . . . in American Indian traditions' (2010: 83).[20] Rather than a hubristic claim of human domination, it is an emphasis on embeddedness in relations that are ontologically prior to categorical distinctions between human and nonhuman, between matter and idea, between individual and social. When such categories are understood as denoting static sets of entities of different kinds, they function as misplaced abstractions in Whitehead's sense, since all entities are manners of relational processes and events.

A metaphysics of process as explored in the first part of this book is necessary for understanding how patterns or styles transfer across ecological registers (material, psychic, social). Manifestation of these patterns vary, but Guattari observes the transference of pathological patterns while challenging the habit of assuming their ontological source in one register (whether the social, the material or the psychic). How do we understand this transfer? Is Guattari just making an analogy? Or is there something real about the transfer of patterns?

If we presume a substantial or essentialist metaphysics and the bifurcation of nature, Guattari's observations can only be metaphorical. But Guattari means them to be more than this. Consider the monocultural production of GMO large-scale industrial crops replacing small-scale subsistence farming adapted to local conditions as a material practice that mirrors what Lorraine Code has aptly termed 'monoculture epistemology' with its emphasis on validation of *a sole model of knowledge production* (2006: 8–9). To be sure, this production is motivated by material demands within an industrialised economic system. But it is also an expression of conceptual habits (valorisation of perceived efficiency, for example; lack of attention to *qualitative* specificity, quantity as a single rubric – how do we make the most food, not the best or most qualitatively rich – and so on) and social and economic structures (profit as the overriding motivation). We hear arguments that this scale of production is the only means necessary to meet demand, but it is also possible to understand that this demand itself is produced by social and psychic habits and patterns.[21] (The emphasis on degree is important. Humans have to eat; the question is how.)

Guattari's emphasis on the psychic is in the context of what Bateson has called an 'ecology of bad ideas'[22] – the way that psychic habits and patterns of feeling and attention are reproduced and perpetuated in manners that sustain pernicious material or social practice, even as these material and social practices manifest in rising psychic illness. This is *not* a unidirectional causal claim. It is *not only* a matter of psychic habits. But the psychic habits matter – both as expressions of webs of dysfunctions and also as sustainers or producers of those webs: 'Just as monstrous and mutant algae invade the lagoon of Venice, so our television screens are populated, saturated, by "degenerate" images and statements' (Guattari 2008: 29/34).

Changes in psychic ecologies alone are not sufficient for transforming 'mutant algaes'. But the very notion of a change in psychic ecology 'alone' is inconsistent with the basic premise of the three ecologies, where psychic ecology is inextricable from material and social. Because of their constitutive relationality, change can never be isolated to only one register. Any change in one has relations to changes in other domains, even if these resist any definitive hierarchical chain of causality.[23] If there is at least some relation between an ecology of bad ideas and the behaviours and systems that drive material exploitation, oppression and abuse, the question is how to engender an ecology of good ideas; that is, how to create positive feedback loops between the social, material and psychic rather than negative ones. One could work on such feedback loops from different points of entry. No one entry (the social, the psychic, the material) is sufficient, but all are necessary. Such a project must identify pervasive features of the ecology of bad ideas to think how they can be challenged and transformed. Guattari's interest in the processes that produce contemporary subjectivity is always oriented by this premise: 'we need new social and aesthetic practices, new practices of the Self in relation to the other, to the foreign, the strange' (2008: 45).

Without further attention to metaphysics this call appears solely idealistic. This is why the frequency of Guattari's appeals to process is noteworthy. Guattari describes process as 'oppose[d] to system or to structure [in] striv[ing] to capture *existence in the very act of its constitution*' (2008: 30, emphasis added). These appeals include reference to 'a logic of intensities or eco-logic [that] is concerned . . . with the movement and intensity of evolutive processes'; a characterisation of mental ecologies as 'primary process[es]' that are 'pre-personal' and 'pre-objectal'; and references to 'processes of singularization' and a 'processual semiotics' for considering how media condition the production of subjectivity (Guattari 2008: 30, 36, 40). Each case stresses the extent to which subjectivity is produced through processes that operate at pre-personal levels. This is not an evacuation of agency as such or a strict determinism, but it does understand that subjectivity can never posit itself as fully given independent of the processes that it emerges out of. The full depth of this point is neither developmental nor psychological, it is metaphysical. That is, it links to an understanding of reality as processual, dynamic and potentially creative.

In order to think this intersection between Guattari's call for transformation and more developed details of a process conception of subjectivity, we have to consider what Guattari calls an 'existential territory' (Guattari

2008: 23).[24] Guattari pairs this concept with what he calls 'incorporeal Universes' (2008: 85). Incorporeal universes function similarly to Whitehead's eternal objects in providing a conceptual rudder of potentials that become actualised in finite and concrete situations. Where incorporeal universes are 'non-dimensioned, non-coordinated, trans-sensible, and infinite [in the sense of limitless]', an existential territory, by contrast, is 'singular . . . sensible, and finite' (2008: 85). It is always the unique and particular territory that it is, even as it shares features and processes across its boundaries.

An existential territory emerges in the overlap between the psychic, the social and the material. Each of these are constituted by ontological processes that cross the categorical (fixed and coordinated) references of conventional representations. The existential territory in this sense is not simply a description of one's lived environment as one perceives it, though it would include this perception as a relevant feature of that territory. Rather the existential territory is the actualisation of a living subjectivity that includes but also exceeds the conscious or phenomenological to include sub-representative affective processes, non-conscious physical processes and relations beyond the cognitive. This means that an existential territory need not be only human, though Guattari's analysis is primarily concerned with human existential territories. Nor is it reducible exclusively to the spatial or material territory in which one lives, since it also includes constitutive psychic processes as well as social formations, obligations, identities and pressures, all of which can differ radically within the same spatial territory. Think of the complex interplay of tastes, drives, styles, structures, place, history, age and material forces that come together to constitute 'where' one lives, the modes of one's 'life-style', and so on.[25]

As locations of reference, neither social relations, psychic processes nor material and environmental milieus function as independent closed systems; rather, they are sustained through processes contingently liable to divergent qualitative outcomes. Characterising a 'common principle' for all three, Guattari writes:

> each of the existential territories with which they confront us is not given as an in-itself, closed in on itself, but instead as a for-itself that is precarious, finite, finitized, singular, singularized, capable of bifurcating into stratified and deathly repetitions or of opening up processually from a praxis that enables it to be made habitable by a human project. (2008: 35)

An existential territory is limited by specific conditions and situations, which is why it is finite and singular and liable to breakdown and dissolution, hence its precariousness. But within these limits, different responses are possible – processes can mitigate against chance through an insistence on static repetition, a repetition that tends towards decreasing degrees of novelty, freedom and life, or they can continually reinvent themselves so as to make further novelty, freedoms and life possible. One strategy attempts to preserve perceived identity at the expense of difference or change, the second adapts a more creative strategy. This tension is correlative with Whitehead's analysis of the two poles of the solutions for the problem of survival.

What Whitehead and Deleuze offer is a more philosophically developed underlying metaphysics that enables Guattari's applied discussion of human subjectivity. The questions now become: given this metaphysical understanding, how does attention impact existential territories? Can we conceive of, and even develop, practices of attention that are not the foregone results of the four operative 'semiotic regimes' for the production of subjectivity: the economic ('monetary, financial, accounting . . . mechanisms'), the juridical ('title deeds, legislation, and regulation of all kinds'), the techno-scientific ('plans, diagrams, programmes, studies, research, etc.') and 'semiotics of subjectification, of which some coincide with those already mentioned, but to which we should add . . . those relating to architecture, town planning, public facilities, etc.' (2008: 32)? What might these look like? One way of thinking stakes of attention when its function is ontologically creative is to diagnose how prevailing axiological logics structure attention. Most notable is the fundamental function of a logic of equivalence to operations of IWC.

The infiltration of equivalence: from exchange to axiology

At the heart of ecology, whether in its scientific function or philosophical concept, is the relation between unity and diversity. In ecological sciences, this manifests in debates over the relationship of diversity and stability. While it is a core intuition of many ecologists that 'diversity begets stability', because of 'the multiplicity of possible definitions of "diversity" and "stability" . . . there are probably no better instances of formalization indeterminacy in any scientific context' (Sarkar 2005). Indeed, there is no consensus amongst biologists about how to quantify biodiversity, not to mention stability.[26] These debates are not only academic, rather how we understand the relationship between diversity and stability has significant normative implications for approaches to conservation. Such practical questions resonate with ecology's philosophical formulations as well. Is it possible to understand a relational ecosystem without collapsing significant differences into a single rubric, metric or measure?

Guattari observes that 'ecological disequilibrium' manifests in 'progressively deteriorating' modes of 'both individual or collective life' (2008: 19). Such deterioration is demonstrated in rising pathologies at social and psychic levels: 'oppressive marginalization, loneliness, boredom, anxiety, neurosis' (2008: 20) and the 'propagation of social and mental neo-archaisms' evident in resurgences of religious or military fundamentalisms (2008: 42). If this might be described as collapse of stability, how we characterise it with regard to diversity is less clear. Guattari shows that fixation on a static logic of identity can generate negative feedback looks that intensify processes of such destabilisations. When diversity or difference is seen as a source of destabilisation, a response is to reject it or shut it down in some way. But this leads to further cycles of isolation and domination.

Another way of approaching this dynamic involves closer thought about how we conceptualise stability. When stability is construed too narrowly as *repetition of the same*, it will be unable to conceive of difference as an opportunity for evolution of the whole. It is necessary for both diversity and stability to be conceptualised differently. Stability cannot be conflated with predictability, static identity or homogeneity. Diversity cannot be conceptualised through logics of oppositional identity or 'us v. them' conflict. As Guattari states 'Individuals must become both more united and increasingly different' (2008: 45).

The extent to which such a call appears fatally contradictory reveals how deeply habits of thought are governed by a logic of equivalence intrinsic to the production of predictable subjectivity under capitalism. If Marx understands equivalence as a logical necessity for general exchange, Guattari shows how it morphs from this operational role into a fundamental value that far exceeds exchange in the traditional sense. Equivalence inhibits or reduces the axiological in functioning as the master value to ensure that qualitative difference be put into a single rubric.

Marx identifies *equivalence* as a logical necessity for establishing a way of exchanging commodities with different 'use-values'.[27] Equivalence arises as a means of translating between quantity and quality, seeking to render the qualitatively variability of use-value into a single quantitative form. As Marx says, 'every useful thing, for example, iron, paper, etc., may be looked at from the two points of view of quality and quantity . . . the usefulness of a thing makes it a use-value' (1994: 220). This usefulness varies according to context and is historically developed: 'The discovery of . . . the manifold uses of things is the work of history' (Marx 1994: 220). But given the variability of uses, purposes and contexts, use-value is not consistent enough to establish regular procedures of exchange. For this reason, Marx contrasts it with 'exchange-value' which 'appears first as the quantitative relation in which use-values of one kind exchange for use-values of another kind' (1994: 221). Here the principle of equivalence becomes fundamental. Equivalence becomes the form of value shaped by an operative generality that overrides qualitative difference: 'The universal equivalent form is *a form of value in general*' (Marx 1992: 162, emphasis added).

Marx is interested in this move from the qualitative variability of use-value to equivalence as exchange-value because of how it enables capitalist accumulation and exploitation. One way it does this is by separating the concrete material labour that inheres in qualitatively different commodities and making of them one abstract measure of value – exchange-value, governed by a principle of equivalence. Given historical developments of capitalism since Marx, my interest is less in terms of material practices of labour, and more in terms of the production of constrained subjectivities. That is, equivalence moves from a principle of exchange converting qualitative into quantitative to a powerful norm that infiltrates habits of attention. In this sense, equivalence renders alternative, non-economic forms of value as *prima facie* incoherent. In so doing, equivalence, as a principle, has effects on the potentials of sensibility. It dulls and impoverishes manners of attention.

For both Guattari and Jean-Luc Nancy, this dulling of attention is a crucial aspect of the power of contemporary capitalism *to reproduce itself despite increasingly visible negative ecological consequences*. It is a matter of capturing alternative forms of value before they can even emerge. Guattari writes: 'What condemns the capitalist value system is that it is characterized by its general equivalence, which flattens out all other forms of value, alienating them in its hegemony' (2008: 43). Consider for example Nestlé's chairman Peter Brabeck-Letmarthe's insistence that water be given a 'market value' so that we are aware that it 'has a price' (Sassen 2014: 192), as if this were the only form that value could take. Through the hegemony of equivalence, there is an infiltration of subjectivity such that a capitalist consciousness becomes simply 'natural common sense': 'the capitalist process is as much . . . an enterprise of the production of subjectivity as it is of the production of goods' (Massumi 2018: 82). This is not to deny the importance of material practices, but it is to observe the way in which capitalism functions by 'capturing the future' as Massumi puts it: 'the capture of the future is the capture of potential, change, becoming' (2018: 38). Equivalence as form of value ensures that change can only be understood through its manner of producing the same – profit.

Similarly, Nancy observes, 'the regime of general equivalence henceforth virtually absorbs, well beyond the monetary or financial sphere but thanks to it and with regard to it, all the spheres of existence of humans, and along with them all things that exist' (2015: 5). It is not just money that Nancy is talking about, it is *the form of equivalence as an invisible normative principle*. In addition to its economic role, such a form is incarnated in the ubiquity of digitalised information or data, an equivalence that codifies everything into translatable series of binary 1s and 0s. Donna Haraway presciently described this in 1984 as 'the translation of the world into a problem of coding' (2001: 302). Though coding appears as a neutral technological procedure, it reinforces the form of equivalence by operationalising what Nancy calls the 'limitless interchangeability of forces, products, agents, or actors, meanings or values' (2015: 5). The issue involves characterising connection without reducing it to equivalence. Betraying once again their common root, ecological thinking is inextricable from economy. It is a question of modes of connecting and relating that are not captured by equivalence. Put it like this: can we connect, can we relate, to that with which we do not render transparent through a single form of value?

This question remains speculative and existential. But it can only be concretised if we come to better realise the powerful effects of equivalence. These effects have become only more entrenched in the transformation to a digitalised economy where 'data is the new oil' (Foer 2017: 186). Information, as a quantitative tool, is governed by a principle of equivalence where everything is transparent in form. Everything is data. Or conversely, if there exists some 'thing' (feeling, experience, quality, sense?) that cannot be translated into this matrix of equivalence and exchange, it is irrelevant and rendered valueless. Hence the irony of the well-known Mastercard ad campaign with its appeal to the 'priceless' that actually reinforces the converse: all values are ultimately financial.[28]

The principle of general equivalence controls increasing layers of interconnectivity mediated through communicative technology. It is as if the very opposite of Guattari's call is occurring. In becoming increasingly connected, there is no unity and less difference. Is this a necessary result?[29] How does the form of the principle of equivalence impoverish different existential or qualitative possibilities inherent in new technologies? Mark Hansen draws on Whitehead's metaphysics to argue that twenty-first-century media produces a 'surplus of sensibility' through its harvesting of data operative below thresholds of conscious awareness, both in terms of the micro-sensory automatic surveillance and the tracking built into our prosthetic digital technologies.[30] Hansen explores how this surplus could give rise to other modes of behaviour, in effect, to other ways of living experience even as the procedures themselves challenge the centrality of the phenomenological subject. Whitehead's metaphysics, with its emphasis on the ontological primacy of 'feeling' offers a way of thinking this combination of a production of sensibility that is not necessarily linked to the subject. The question though has to do with how this production is used. Currently, it is governed by what Hansen calls 'the narrow instrumentality of capitalist cultural industries' (2015a: 66). If such a surplus is to succeed in challenging the repetition of this instrumentality, it must evade the form of equivalence. Moreover, while Hansen is right to suggest that the precognitive or sup-perceptual nature of the data-harvesting challenges the primacy of the phenomenological subject and that a Whiteheadean process metaphysics can help to render this coherent, this does not answer the existential question. It is one thing to 'know' in a conceptual sense that one's subjectivity is constrained, encouraged and otherwise shaped by forces that exceed conscious awareness, it is another thing to think how to incorporate this notion into one's habits of sensibility and attention.

One step towards such incorporation is becoming more aware of the procedures by which one is encouraged to remain blind to these forces. Indeed, any such transformative mode of sensibility and differently attuned sensibilities must reckon with powerful procedures of capture that seek to reincorporate them into the fundamental logic of IWC. This exposes a second, *temporal*, dimension to the form of equivalence. This temporal dimension is intensified in the context of big data and digital technology where equivalence as a form of value is aligned with predict-ability. Equivalence is a tool for perpetuating exchange into the future by making behaviour predictable. Equivalence-prediction are two modes of the same logic. When Hansen claims that Internet searches 'automatically produce surplus value', he is referring to this temporal function (2015a: 187–9). The value of an individual search is more than just its meaning in the present because it feeds into the algorithmic data-mining on which the search engine depends and becomes data for future prediction. The more repeated terms in a net aggregate, the stronger the probability that they repeat in the future. What you see is what you get.

Data, the new oil, are the mode by which capitalism predicts what you will click on and what you will buy. It is not data as such that are the commodity, but rather data as means *to capture attention*. Indeed, Hansen poses the problem in terms of the 'contemporary exploitation of the time of attention by capital' (2015a: 189). Attention becomes a lim-ited, exploitable, resource, and its capture the newest mode by which capi-talism continues. Indeed, this exploitation is all the more pervasive insofar as digital technologies operate at precognitive levels to nudge conscious behaviour without your being aware of it.

What Hansen calls the 'imperative driving today's data industries . . . to create closed data loops' (2015a: 200) is entirely oriented by the logic of equivalence in the interests of prediction. Indeed, equivalence-prediction is *the form of closure*. It ensures self-fulfilling prophecies by disregarding possibilities that do not already appear in forms that can be assimilated into equivalence. It is no longer just an operation for the exchange of *commodities* but rather structures the translation of living experience (feeling) into regulated and transferable meaning. Value inflects perceived meaning. The limitation of value to one form is the diminishment of meaning.

When equivalence-prediction is the principle of translation of the lived to the intelligible, it reduces the sensible and qualitative to mere decoration extraneous to the 'real' meaning of the material-economic. In this way, in addition to the criterion of profit, equivalence-prediction diminishes or otherwise inoculates the capacity for encountering surprise or the unknown in any robust sense. We still may live the possibility of trivial variation, but equivalence-prediction as dominant form of value enables an assumption that meaning is, at a basic level, fixed. While we may not know what *is*, we do know that whatever is, its value is already to be found through the rubric of equivalence: 'there is no longer strictly speaking any confrontation with the other since it is absolutely the same confronting the same' (Nancy 2015: 22).

This is the second mode by which equivalence-prediction impoverishes attention. On the one hand, there is the capture of attention as resource to enable predictable behaviour. This can be called a passive diminishment insofar as it happens *to you*. But equivalence-prediction also has an *active* effect on attention through its diminishment of encounter. *When we cannot be surprised, there is no need to pay attention.* Even if we don't know the details, we already know what matters because there is only one form of mattering. This is why homogenising and standardising in the name of efficiency also reinforces increasingly closed loops that deny the need to *pay attention* to qualitative singularity. If such singularity is even granted as possible, it is by definition value-less, since 'the value of any value is its equivalence' (Nancy 2015: 5). When there is only one value that matters, attention atrophies.

Guattari, drawing on his work with Deleuze, an implicit process metaphysics, and his background in psychoanalysis, explores how this logic infiltrates the production of subjectivity such that the subject understands it as self-evident, as part of a more or less freely chosen value system rather than an imposition of control. This is a complicated story which has been told from many angles, perhaps most notably in the Foucaultian terrain of the biopolitical and the disciplinary or by Deleuze and Guattari in *Anti-Oedipus*. It is easy to see how equivalence serves the interests of capital: subjects whose desires and thoughts are predictably similar better serve the needs of a machine whose only criteria for value is profit. In this sense, while Guattari's claim that a 'global market destroys specific value systems and puts them on the same plane of equivalence: material assets, cultural

assets, wildlife areas, and so on' (Guattari 2008: 20) is increasingly undeniable, such observation is not in-itself enough, precisely because of the
extent to which equivalence is internalised and normalised.

How is this accomplished? We can attribute some of this to the force
of collective normativity and social discipline where an entire apparatus of the 'semiotic of subjectification' operates according to standardisa-
tion, homogenisation and normalisation. Such procedures produce and
perpetuate mass subjects with predictably similar desires. To the extent
that deviations occur, these are reincorporated into the prevailing logic of
prediction and consumption through finding a way to make deviations
profitable. Yesterday's subversive revolutionary is today's fashion icon. In
this sense, there is no subversion of the system that is not assimilated into
the logic of equivalence-prediction. Indeed, a certain amount of subversion *is even necessary* to keep creation of new modes of consumption from
stagnating. But just a certain amount and only in certain ways. In this way:

> A capitalistic subjectivity is engendered through operators of all types and
> sizes and is manufactured to protect existence from any intrusion of events
> that might disturb or disrupt public opinion. *It demands that all singularity
> must be either evaded or crushed in specialist apparatuses and frames of refer-
> ence.* (Guattari 2008: 33, emphasis added)

Given the extent to which IWC can incorporate stylistic difference and
make of it a new commodity, the challenge is significant. Singularity here
must be intended in an ontologically creative sense, or else it is at risk of
faddism alone. This is why Guattari is so concerned to interrogate how
IWC operates within subjectivity, observing that it 'tends increasingly to
decentre its sites of power, moving away from structures producing goods
and services towards structures producing signs, syntax and – in particular,
through the control which it exercises over the media, advertising polls,
etc. – subjectivity' (2008: 32). As we have seen, Brian Massumi takes this
Guattarian insight to the extreme in claiming that the production of subjectivity is the means by which capitalism captures the future (2018: 82).
This is significantly demonstrated in how it functions through financial
markets, which in many cases are nearly entirely speculative, dematerial-
ised or liquidified. Here the temporal dimension of equivalence becomes
paramount insofar as financial markets are inextricable from prediction.
Moreover, though this is denied in the explicit, Massumi observes that
financial markets, 'run more on affect and intensification than on underlying economic "fundamentals"' (2018: 133).

Fear, confidence, and so on, become drivers of financial speculation.
Capitalist power, at the same time that it has become increasingly diffuse
in the spatial sites of material production, extends its influence over 'the

whole social, economic, and cultural life of the planet' especially 'in 'inten-sion', by infiltrating the most unconscious subjective strata', the affective (Guattari 2008: 33). The power of advertising is only a surface manifes-tation of this infiltration which extends into the very grain of subjectiv-ity, shaping desires and dreams. 'The worlds of childhood, love, art', as unpredictable sources of potential difference, are anaesthetised through a proliferation of conforming narratives, images and ideology 'hook[ed] up to ideas of race, nation, the professional work-force, competitive sports, a dominating masculinity, mass-media celebrity' (Guattari 2008: 34). In what Guattari aptly calls 'the mass media imaginary' (2008: 35), 'capitalist subjectivity seeks to gain power by controlling and neutralizing the maxi-mum number of existential refrains' (2008: 34). Despite celebration of the freedom of the individual consumer, the actual situation entails increas-ing homogenisation and commodification of those who 'think differently' into icons of consumption.

It is easy to pretend that denunciation of such homogenisation alone is sufficient for escaping these effects. Nothing could be further from the case. First, as Isabelle Stengers observes, 'However justified, denunciation fabricates a division between those who know and those who are duped by appearances. Worse, the knowledge that it produces has no other effect than to attribute even more power to capitalism' (2015: 74). This contrib-utes to a sense that the problem is a false consciousness whose antidote would be the uncovering of a real or authentic consciousness purified of the alienating or otherwise corrupting influences of the mass media, advertis-ing, and so on. This fails to understand that subjectivity *is always produced through processes* that are ongoing and incomplete. It is thus too broad in positing an essential difference between a primary authentic consciousness and a false consciousness. When posited as such a complete entity, con-sciousness is an overly broad abstraction. A subject is not one consciousness as such, it is a series (Whitehead's 'historic route of occasions') of moments that share enough of a pattern so as to achieve a coherence.

Indeed, part of the function of equivalence is to reify broad abstractions and make of them static things. This would apply to references to capital-ism as well. While certainly this names something important and as such is useful in certain contexts, it also can slip into a false reification of what is, in effect, a system constituted through ongoing actions, behaviours and modes of production and exchange. This is not to etherealise the material. Capitalist manners of exchange, systems of production, modes of finance, and so on, are powerfully real and shape actual occasions. But the com-plexity of these systemic functions, coupled with ideological patterns and habits of attention and behaviour, mean that their dissolution or transfor-mation will be similarly complex and multivalent.

A processual approach to subjectivity, as a metaphysical claim, means that the production of subjectivity is not unique to life under capitalism. There is no originary *a priori* authentic process of subjectification that capitalism covers over. A processual conception of subjectivity understands the repetition of affective patterns to be constitutive of the emergence of a coherent subject as such. This is the case regardless of the economic or political system within which these patterns manifest. However, this is not to posit an essential separation, quite the contrary. Repetition is at the heart of the emergence of a living subjectivity, and repetition is structured, in part, by societal, cultural and economic forces. In particular, when repetition is structured by operations of equivalence, it is easily enlisted in perpetuating dulling or even compulsive behaviours. It is always a question of quality and consequences. Whitehead's analysis of societies provides a basic conceptual polarity: in what sense do repetitions feed back into the future? Do they lead to enriched subjectivities, with more intense or broader 'bandwidths' of information? Or do they lead to narrower and narrower scopes of awareness (a numbing form of mass hypnosis)?[31]

It is thus too simple to denounce repetition as such, precisely because any coherent 'existential territory' is in part created through 'autoreferential effects'. Such auto-referential effects operate at a number of levels, including: 'the repetitive symptom, the prayer, the ritual of the "session", the order-word, the emblem, the refrain, the facialitary crystallization of the celebrity . . . [all of which] initiates the production of a partial subjectivity' (Guattari 2008: 37). Though Guattari stresses the way that repetition can become pathological, it is very much a question of choosing your symptom – becoming your refrain, if you will, and learning to attune to its own potentials for thresholds for transformation. This cannot be a one-size-fits-all diagnosis in its content. The issue is not to somehow escape procedures of auto-reference as such, but rather to think how differently attuned attention might orient such 'refrains' towards alternative qualities and consequences. This is why it is a question of the 'aethetico-existential *affectiveness*' (Guattari 2008: 37, emphasis added) of the repeated forms of auto-reference such that both Massumi and Guattari identify the need for a 'non-normative ethico-aesthetic' (Massumi 2018: 95; Guattari 2000: passim) in transforming the habits of attention by which capitalist subjectivity perpetuates itself.

For Guattari, 'the crucial objective is to grasp the a-signifying points of rupture' (2008: 37), the places where recognition breaks down, where gaps or disruptions to procedures of equivalence may allow for the emergence of something different or unexpected. Given the extent to which it necessarily challenges the smooth functioning of identity, such a rupture is perilous. What would it mean for attention to produce an existential territory attuned to such disruptions or differences? If Guattari links this to what he calls the 'principle' of mental ecology, how is such a principle informed by the metaphysical results of the first half of this text?

Attention as ontological and counter habits

Since subjects emerge through affective or feeling patterns, then in the interest of challenging prevailing attunements easily enlisted by capitalist subjectivity, alternative modes of sensibility must be cultivated. This claim can appear simple, even trite. On the one hand, it is far more easily said than done. On the other hand, appeal to alternative modes of sensibility risk becoming what Murray Bookchin criticises as merely 'lifestyle anarchy' if we do not attend in greater depth to how attention functions ontologically and what is really meant by an alternative mode of sensibility.[32] Finally, characterising such attunement as ecological requires considering what this means without presuming it automatically maps to conceptions of the environmental. Indeed, part of the promise of an ecological attunement, if also part of its challenge, is its potential emergence in a variety of settings, from the more traditionally 'natural' to the heavily constructed or mediated. If certain milieus may be more conducive to the emergence of alternative attunements then others, and if conditions of pervasive violence or trauma pose particular challenges, in principle, understanding attention as ontologically creative can encourage a shift in attunement towards the ecological *in any context*.

Understanding this potential requires tackling a significant paradox. How can attention be creative when its most pervasive operation is commonly oriented towards the expected, towards the 'object of recognition', as Deleuze would say? It also requires further discussion about why such an attunement can be called ecological. Though 'ecological' often carries a normative valence, the claim here is thornier – an ecological attunement cannot consistently present definitive prescriptive claims.

This is connected to a basic tension in Guattari's admonition that 'Individuals must become both more united and increasingly different' (Guattari 2008: 45). In response to how a logic of equivalence sees everything through a single lens, the creation of subjectivities that singularly activate an existential territory through values resistant to equivalence is necessary. But at the same time, for such activation to be sustainable, it must forge new forms of alliance or connection. The chief tension is between this need for new forms of connection and creations of 'singularity' (Guattari 2008: 33). How does singularity connect? This question is inherently ecological, which is to say relational. Can we conceptualise connection without presuming a single set of values or parameters?

Guattari has this in mind in stating: 'Rather than looking for a stupefying and infantilizing consensus, it will be a question in the future of cultivating a *dissensus*' (Guattari 2008: 33). This point is easily co-opted when *dissensus* is only a reactionary mode of resisting whatever dominant model is taken as norm.[33] Orienting through resistance to the norm alone perpetuates the *form of normativity* as such. What must be conceptualised instead is the emergence of value in a manner that does not reduce to prescriptive universal judgement. Such emergence is ecological insofar as it does not pre-exist the relations through which it is constituted. How can we think creative emergence in an existential key without falling into a prescriptive logic bound up with single-register planes of dominance? If Guattari calls for 'reappropriating Universes of value, so that processes of singularization can rediscover their consistency' (2008: 45), he does not, indeed cannot, substantively describe what such re-appropriation looks like prior to any singularisation in question. Nevertheless, he does make the positive claim that transforming ecological destruction requires re-energising human creative potential: 'The reconquest of a degree of creative autonomy in one particular domain encourages conquests in other domains – the catalyst for a gradual reforging and renewal of humanity's confidence in itself starting at the most miniscule level' (2008: 45).

Starting at the most miniscule level means beginning with attention in the micro-grain of each moment. But simply urging attention is insufficient if not accompanied by reflection on a series of conceptual tensions inherent in translating metaphysical results into existential habits. The first is the problem of ecological value: how can we think of value as a creation of events that does not antedate the relations in question? Attention is a form of activity that creates value insofar as it is a valuing. And yet, how can we orient attention without some preconceived notion of value – some notion of what *should* be attended to? The conceptual task inheres in transforming the substantialist components in those concepts: attention and will. Attending is not a passive or accidental predicate affixed to a presumed subject, but rather continuing actions by which subjects are sustained. It is not a question of *will* so much as one of *willingness* and *awareness*. Changes of attention are changes in relation – this is a tautology, provided it is tempered with the understanding that relations are by degrees. Attention makes things and does things – what do 'you' want to do?

Where concepts of 'the' will presuppose isolated autonomy, willingness understands the conspiratorial nature of attending as unavoidable. This remains a question of degrees. Undoing metaphysical autonomy does not deny choice, on the contrary, it makes choices matter more. But choices of attention are never total or unilateral exercises of full mastery (this is consistent with what cognitive science tells us. Our explanations of why we have chosen what we choose are largely retroactive.). Attending as conspiring requires *willingness* to relinquish control of the result. Choices always occur within a pervasive paradox – you cannot fully control conditions *or* outcomes and yet both are still in some sense the result of ('your') choice. The key shift though is that choice is not by fiat, it is always a matter of collaboration.

The problem keeps repeating. Metaphysical antinomy becomes ecological antinomy: how to affirm both qualitative singularity and interdependent relatedness? How to think singularity that is not a form of separation? Conversely, how to think connection or relation not governed by equivalence? It is a matter of understanding where to place the stress in what Whitehead calls the 'ideal opposites' of 'permanence and flux' (PR: 338). Is a singularity that which arrests the flux? Is relation that which generates the dissolution of singularity? Or is it rather the very condition of singularity? Whitehead shifts the frequent scope of these 'ideal opposites' by both destabilising those entities typically taken to represent permanence and, at the same time, eternalising the ephemeral. Referring to Michelangelo's *Day* and *Night, Evening* and *Dawn* statues in the Medici chapel of Florence,[34] Whitehead writes

> perfect realization is not merely the exemplification of what in abstraction is timeless. It does more: *it implants timelessness on what in its essence is passing.* The perfect moment is fadeless in the lapse of time. Time has then lost its character of 'perpetual perishing'; it becomes the 'moving image of eternity.' (PR: 338, emphasis added)[35]

The timelessness is not implanted onto the work of art as artefact or entity, rather it is the moment of the encounter – the collaboration between attention, colour, light, form, and so on, that engenders such singular moments. In this sense, the inversion is complete – all artefacts are ultimately temporary, what remains is the timelessness of the singular moments. And yet, such moments, which Whitehead does not hesitate to characterise as 'immortal', cannot of course be captured or retained in any substantive way because they are 'perpetually perishing'. We are 'faced by the paradox' from which there is no escape. This paradox is not only, or even primarily, a conceptual problem. It is rather a living, which is to say felt, experience. The task for thought is to construct a metaphysics that does not explain away this paradox.

The conceptual antinomy functions as a double bind within which there can be no absolute salvation. And yet, the constraint it enacts, the impossibility of a totalitarian possession of permanence, or, put conversely, the sole realisation of permanence in the ephemeral moment, also makes possible what I call *ecological attunement*. Such attunement is not advanced *as a solution* to loss, to the diminishment of life, to ecological calamity. In its humility, such attunement is clearly insufficient to the scale of interlocking systemic crises driving such destruction. And yet, what else could it mean to begin at the most miniscule level? Accordingly, to learn to attend differently, even if insufficient, is nonetheless vitally necessary. The question is not how to achieve permanence construed as refuge from the world's disarray; rather, it is how to select or emphasise what one wants to preserve or pass on *in the ephemeral experience of the moment*. Ecological attunement is attuned to this dual sense of permanence and passage – where the stakes of each moment's attendings are inextricable from creation of the future even as they inevitably, necessarily, fail to inoculate that future against its own peril.

Ecological attunement, *as concept*, intensifies rather than eradicates three insoluble tensions that equivalence covers over: (1) the creation of novel value or meaning that is retroactively construed as having already been there; (2) the necessity *and* practical difficulty of *multi-scalar attention* and (3) the art of choosing 'refrains' you inherit. Each of these, in varying degrees, *follows a concept of attending as relational and actively creative rather than unidirectional and purely receptive or passive*.

Equivalence proposes to predetermine possibilities of value or meaning by controlling the form of all value as such. This denies the possibility of novel value or unanticipated meaning. In this sense, it is incapable of thinking a robust conception of relation where *the quality of the relation* is creative of new forms of value or meaning. The particular relata in question do not matter since the form of relation is always the same. When attending is instead part of the ontological creation of each actual occasion, even as it passes, then this creation opens to alternative emergences of meaning or value. It is true that there are some limits to these alternatives, but also that these limits are not deducible only from contingencies of experience. In the language of Deleuze or Whitehead, neither the eternal objects nor the virtual are exhausted by what has happened. But this is also why equivalence functions to impoverish reality insofar as it inhibits the potential of novelty.

Dangers remain. Attending risks becoming assimilation if it is oriented by a presumption that purposes are shared. This is a failure of attention because it assimilates rather than collaborates. An ecological attunement involves accepting the necessity of encounters that are not synthesised through a shared sense of purpose or consensus as to their meaning in relation to stable image of the whole. Rather, events of relation and encounter are creative of values later used to characterise their meaning.

Isabelle Stengers has an excellent sense of this problem in her articulation of what she calls an 'ecology of practices' that does not fixate on a single universal value as the only viable ordering of a multiplicity. Such an ecology of practices departs from the (mainstream) metaphysical tradition's dream of '"delocalized" concepts, which guarantee the ability to travel anywhere and to be at home wherever one happens to be' (Stengers 2010: 62). Such universal delocalisation is motivated to escape the need for attention to constraints generated by local conditions. Stengers calls for a different form of delocalisation, one that following Deleuze can be called 'nomadic' rather than 'sedentary' (DR: 36–7).[36] Indeed a universalistic delocalisation brings all places or events to order under one measure. It is in this sense an expansion or incorporation into a pre-existent order. Delocalisation eradicates the possibility of a novel value of the local by ensuring that this value can only be measured according to the pre-existent. By contrast, Stengers is interested in a delocalisation that is horizontal rather than hierarchical. Such a delocalisation would be genuinely relational because it would destabilise any rubric deemed capable of judging before the event of relation. Such a delocalisation in Stengers's sense

> bring[s] into existence the experience of here *and* there, the experience of a here that, by its very topology, affirms the existence of a there, and affirms it in a way that excludes any nostalgia for the possibility of erasing differences, of creating an all-purpose experience. (2010: 62)

Such delocalisation does not 'reterritorialize' (in Deleuze and Guattari's language) on a permanent absolute determination, but rather intensifies the need for attention by denying the predetermined adequacy of any concept deemed universally sufficient. This destabilising of any assumed universal is motivated by a robust concept of relational encounter where creation is a potential result that does not depend on intention.

Unintentional creation of meanings is crucial for ecological attunement. Stengers proposes concepts of 'obligation' and 'requirement' as heuristics for making sense of how values emerge out of practices without these practices requiring a global sense of shared purpose.[37] She characterises obligation and requirement as constraints, rather than rights, duties or conditions:

> Unlike conditions, which are relative to a given existent that needs to be explained, established, or legitimized, constraint provides no explanation, no foundation, no legitimacy. A constraint must be satisfied, but the way that it is satisfied remains, by definition, an open question. A constraint must be taken into account, but it does not tell us how it should be taken into account. (2010: 43)[38]

As Stengers suggests, 'every living being may be approached in terms of the questions of the *requirements* on which not its survival but also its activity depend . . . and every living being brings into existence *obligations* that qualify what we refer to as its behavior' (2010: 54). Despite Stengers's reference to 'living beings', these concepts are abstractly flexible enough for thinking processes of relation constitutive of complex ecologies, not just organisms and environments. They can be used to describe the interaction between different species, different materials and different physical vectors as well. In this sense, they are metaphysical concepts that 'identify *a priori* neither the user nor the field of use' (Stengers 2010: 61). The only constant is that their use always brings into view a 'here' and a 'there'. This is their 'topology': 'We require something from something or someone. We are obligated by, or are obligated to [something]' (2010: 52).

A topological distinction between a here and a there does not pre-exist the operation in question. In contrast to universal delocalisation, activities of requirement/obligation vary in accordance with the particular existential or ontological field in which the events that they constrain occur. They are not simply applied in the same way every time, but rather serve as orienting guides to analysis of the practical, ontological or epistemic encounter in question.[39] Stengers writes: 'They function as operators intended to make perceptible, in the very way they must be reworked *to earn their relevance in each practical field*, the topological transformations that mark the transition from one field to another' (2010: 61). The physicist is under a different obligation than the farmer, and yet both proceed in relation to requirements and obligations.

To earn relevance is to induce the event of a satisfied requirement or fulfilled obligation. In this event a value is either affirmed (in the case of ordinary satisfactions) or created (in the case of novel achievements). This achievement does not require that all the constituents of the event share the same understanding of the intentions driving it. It is a peculiarity of an event of novel achievement and the creation of what will then be called a value, that this valuation, is 'detected as soon as [it is] produced, and . . . found to be already there, always, as soon as [it is] engendered' (Stengers 2010: 37).[40] The event of a satisfaction is an actualisation that creates a new value. But in this creation, it is retroactively seen that its potential was always already there.

This is what Guattari describes as an 'expressive a-signifying rupture' (2008: 30). Psychotherapeutic work provides an intuitive context for thinking about such ruptures. Therapeutic achievement can be described as growth, but not in the sense of an ordained growth that follows a predictable route. Consider the growth that may occur when a subject struggling with a particular psychic formation is suddenly able to take it apart differently and reconfigure it, such that it no longer bears the same weight as before. Such a rupture is a break from structuring presuppositions, habits or pathological repetitions – it bifurcates a series or pattern into a new direction. It

> summons forth a creative repetition that forges incorporeal objects, abstract machines, and Universes of value *that make their presence felt as though they had been always 'already there', although they are entirely dependent on the existential event that brings them into play.* (Guattari 2008: 30–1, emphasis added)

This nonlinear creation of value is manifest in episodes of renewal or 'quantum leap' that occur when someone experiences a 'freeing up' after a long period of feeling stuck. The problem is evaluated differently on either side of the rupture – its meaning may change or shift, though there is often a consolidation into a new repetition and set of habits, ensuring that narration will retroactively ascribe this new meaning as inchoately already present.[41]

Because such consolidation is retroactive, it is difficult to conceive of its effects on attention. How can we attend to that which has not yet manifested or emerged? Indeed, while we can conceive of *the idea* of symbiosis in naming events of achievement in which agents with different needs and desires interact in mutually enriching ways, such an event, by definition, cannot be *willed* from a single perspective. It does not require agreement upon the purpose of an interaction, nor does it require that both agents describe *the meaning* of the interaction in the same way. Its value emerges at a different order of complexity excessive to the immediate intention of either agent: 'The "symbiotic agreement" is an event, the production of new, immanent modes of existence, and not the recognition of a more powerful interest before which divergent particular interests would have to bow down' (Stengers 2010: 35).

How can posing this at the level of thought respond to the imperative for attention? At the least, it emphasises a requisite encounter: approaching a complexity of divergent purposes and actors as both linked in a series of constraints and yet not necessarily teleologically oriented towards an achievement of universal consensus. As we have discussed previously, while some Whitehead commentators read his claim of beauty along such teleological lines, Whitehead's God remains fundamentally open and in this sense neutral to the qualitative expression that results from the encounters and 'selections' of occasions. Similarly, when Deleuze and Guattari declare that 'the strata are judgments of God', these judgements are not moral, but rather expressions of the consequences of relations:

> To express is always to sing the glory of God. Every stratum is a judgment of God; not only do plants and animals, orchids and wasps, sing or express themselves, but so do rocks and even rivers, every stratified thing on earth. (ATP: 40–1)

These expressions are thus at once individual, but also always occurrent in a wider context of relational encounter. God's judgement is akin to the necessities that follow these encounters but do not determine them. It is always 'if . . . then', with a degree of openness: 'the earth, or the body with organs, constantly eludes that judgment, flees and becomes destratified, decoded, deterritorialized' (ATP: 40).

Coding and decoding, encounter and expression are not predetermined by intentional goals; rather, they are the means by which 'judgments' emerge. The wasp does not *intend* to help the orchid propagate, and yet the wasp-orchid relation is productive of a valuation beyond the survival of either. Such 'reciprocal capture' or 'double articulation' refers to relational processes in which '*identities that coinvent one another* each integrate a reference to the other for their own benefit' (Stengers 2010: 36, emphasis added).[42] The co-invention of identities is not an all-at-once procedure, rather, such a concept of reciprocal capture privileges events or processes of relation over static entities.

It is challenging to consider how to *think* ecologically without reducing the multiplicity of scales, causalities and temporalities operative. Disparate causalities inform unintentional creation of meanings at widely variant scales.[43] But how can we encounter a multi-scalar temporality and spatiality in our habits of attention? Indeed, thinking ecologically means that *the here-now is never just here and now*. Paradoxically, however, attending to the here-now, in the very grain of its complexity detached from prevailing assumptions through which it is filtered, might, in some cases, enable better awareness of the gaps within equivalence. This is because equivalence operates at the level of Whitehead's 'misplaced abstraction' – it replaces the here-now that is an intersection of multi-valent scales of time and space, with a reduced here-now that is disconnected in its abstraction. An ecological attunement then, paradoxically, understands that practices of attention are always in communication beyond the explicit intention or understanding of the present. They operate at the level of a sensibility that is ontologically excessive to the cognitive or representational. Ecological attunement as concept is as much aspirational asymptotic ideal that one fails to fully reach or realise in the present. But to condemn the concept for failing to achieve such completion is to operate within a different conception of subjectivity than the one offered here. Understanding attending as collaborative or relational means its outcomes remain contingent to circumstances and qualities of encounter excessive to any single perspective. An ecological attunement does not rescue us from this contingency.

Indeed, despite the powerful pull to want to make ecological attunement a determinately normative concept, this temptation threatens to override the emergent structure of valuation. This raises numerous questions that can be phrased using Deleuze and Guattari's language of the 'refrain'. Such a 'refrain' is both existential and ontological. As the child in the dark sings under their breath to comfort themselves, as 'radios and television sets are like sound walls [that] mark territories' (ATP: 311) (an example that might now include the ubiquitous headphones so that everyone is a private territory carrying their own media cloud), the question is not to avoid or deconstruct the 'refrain' as such, but rather to *choose your refrains wisely*. Indeed, Deleuze and Guattari understand that the refrain is not merely a subjective process of psychic development – it is an ontological means for achieving sustainable coherence that operates between 'chaos' and the Earth. The refrain is constituted as a means by which different scales are put into contact. Consider the blessing of the meal. It can be automatic, rote, even a certification of received dogma, but it can also be a form of communication that acknowledges different scales in relation – the terrestrial (Earth, rain, wind), the cosmic (Sun) and the personal. That is what a refrain can do: 'Forces of chaos, terrestrial forces, cosmic forces: all of these confront each other and converge in the territorial refrain' (ATP: 312).

Choosing the refrains you inherit, or, more precisely, attending to the elements within the refrain that best express this germinal function can keep ecological attunement from collapsing into an ineffective embrace of some facile whole. The refrain 'involves an activity of selection, elimination, and extraction in order to prevent the interior forces of the earth from being submerged, to enable them to resist or even to take something from chaos across the filter or sieve' (ATP: 311). We are not the creator of refrains, but rather the creations of the refrains we select or endorse. Not all of these are conscious, but the task of attention is seeking to find the manner in which the refrains we inherit best reveal what is already there. Because the refrain 'always carries earth with it', how can we attend to this earth potential (ATP:312)?

This question cannot be answered in a single valence. The earth, the forces of the earth and the cosmic forces of chaos, are by no means always harmonious. (We eat and are eaten). Even the most apparently tranquil or ameliorative refrains are always at risk of sedimenting and becoming *forces of inattention*. How can we tell the difference? How do we choose or select, especially insofar as choice is excessive to intentions? What is being created in our attendings that we are not aware of? It is here that the question of affect and feeling returns, with a double valence. On the one hand, it is precisely in affect or feeling, understood at the level of sensibility (rather than emotion), that we can be closest to germinal processes of the actual occasion. We can, perhaps, sense the direction of this addition to reality. But this sensing is unreliable and divergent in any communicative context.

Notes

1. I do not mean to suggest that Darwin or Richard Dawkins themselves operates with these conflations, but only to observe the way differing ecological theories lend themselves more or less easily to differing social ideologies.
2. It is also too quick to presume one direction of influence – as if the science of ecology can influence popular conceptions of economy, but not vice versa. Worster's 1977 intellectual history *Nature's Economy* explores the influence of bio-economic metaphors in the development of early twentieth century ecological sciences, using a basic disjunction between the 'arcadian' and 'imperial' traditions. The 'Arcadian' emphasises holism and interdependence, whereas the 'Imperial' exercises reason and labouring power to establish a relationship of productivity between nature and the human.
3. For this reason, Stauffer argues that Darwin can be seen as the true founder of ecology (Stauffer 1957: 139).
4. Nicholas Rescher frequently makes this point. See Rescher 1996: 14–17, 81.
5. This aspiration raises ontological questions in forcing a judgement over what conditions are real and hence valid for inclusion. But willingness to think ecology as entangled with metaphysics depends on an image of metaphysics that departs from some structuring suppositions of metaphysics as well. Most notably, as its root eco- suggests, ecology is predisposed towards a study of relations *between* beings or entities rather than a vertical or transcendental study of the relation between beings and Being or Existence and the Existant as in Heidegger or Levinas. Such a predisposition, while germane to the approach I take, is not shared by all metaphysics as such.
6. Sarkar identifies seven different theoretical problems for philosophy of ecology. I am focusing on those relevant to my concerns. Of tangential interest is also the 'partial observability problem' which bespeaks the challenge of isolating variables to confirm theories (Sarkar 2005).
7. Sarkar identifies two sources for this problem, the fact that ecological systems are 'contingent historically' as well as irreducibly 'complex' (Sarkar 2005).
8. Whitehead makes a similar point. See PR: 9–10, FR: 11–27.
9. A central inspiration for this proposal is Gregory Bateson (especially Bateson 2000: 486–96).
10. Though Deleuze states in an interview that both he and Guattari 'have remained Marxists . . . [insofar as they] do not believe in a political philosophy that would not be centered on an analysis of capitalism and its developments' (cited in Smith 2012: 161), he clearly does not mean this in any doctrinaire sense.
11. The tradition of social ecology associated with Murray Bookchin is also relevant to Guattari's efforts to think the relations between social, economic and environmental pathologies. See Bookchin 1998 for a concise introduction.
12. For more on the connection between Whitehead's process philosophy and the Pope's integral ecology, see Griffin 2016.
13. Where the Pope and Guattari most significantly diverge is in their approach to psychic dimensions of an integrated ecology. The Pope is beholden to certain theological commitments that significantly constrain his approach. For all his insight in pointing to

relations between social and material ecologies, the Pope remains humanist insofar as the human soul is presumed an essential, and exceptional, unity.

14. Guattari does not refer to climate change as such, though his analysis is consistent with its salient feature: the implication of human activities on larger planetary systems. While the science of climate change was in many respects established at the time of Guattari's text, it was not yet a prominent theme in discourse.

15. While Dolly the Sheep and Oncomouse are passé now, and GMO foods crowd our supermarkets, less well-known are biorobotic cyborg locusts and cockroaches. See De Looper 2015 and Washington University 2016.

16. Debate remains over the connection between climate change and the Syrian Civil War. While it was initially reported that drought induced by climate change was a direct factor in civil unrest leading to the conflict, recent studies have been more cautious about attributing a direct connection. Nevertheless, this caution does not entail the opposite extreme – a claim of no relation between the 'natural' and the 'cultural' whatsoever. See Fischetti 2015; Gleick 2014; Nature Editorial Board 2018.

17. Discussions of the Anthropocene are profuse. For its genealogy, see Crutzen 2002 or Steffen et al. 2011.

18. In sociology, see Latour 1991; in anthropology, see Descola 2013 or Kohn 2013; in comparative philosophy, see Norton-Smith 2010. It is worth noting that Descola, Kohn and Latour are all influenced by process metaphysicians, specifically Deleuze, Whitehead and, in the case of Kohn, Peirce.

19. On this correlation between a sharp nature/culture demarcation and religious fatalisms and fundamentalisms, see Latour 2017: 184–220.

20. For additional conformation of this point, see Atleo 2004; Cordova 2007.

21. For a critical summary of this argument, see 'The ETC Report' in Pojman and Pojman 2012: 389–409.

22. 'There is an ecology of bad ideas, just as there is an ecology of weeds, and it is characteristic of the system that basic error propagates itself' (Bateson 2000: 484). Bateson refers to the infamous pollution of Lake Erie: '. . . the eco-mental system called Lake Erie is part of your wider environmental system – . . . if Lake Erie is driven insane, its insanity is incorporated in the larger system of your [human] thought and experience' (2000: 484).

23. This appears both in the negative or pathological patterns discussed above, but also entails positive or creative potentials. As Guattari writes, 'the reconquest of a degree of creative autonomy in one particular domain encourages conquests in other domains' (2008: 45).

24. Guattari's 'existential territory' can be read as an existential application of the broader discussion of territorialisation and deterritorialisation in ATP.

25. Thus far, Guattari's 'existential territory' sounds similar to a phenomenological 'life-world'. In Chapter 6, I show why this is not the case and differentiate the two concepts.

26. For a summary of conceptual and practical challenges in the concept of biodiversity, see Morar et al. 2015.

27. Use-value is the 'usefulness of a thing' in the sense of what we can do with it. Importantly, it is linked to 'the physical body of the commodity itself'. Because it is 'only realized in use or in consumption', use-value cannot be accumulated (Marx 1994: 220–1).

28. The first Mastercard advertisement with the 'priceless' slogan ran in 1997. See Rajamannar 2017.

29. In this regard, Guattari's remarks about the possibility of a 'technological evolution of the media' used for 'noncapitalist goals' in particular through its being 'reappropriated by a multitude of subject-groups capable of redirecting its singularization' (2008: 40–1) appear unduly optimistic and naïve. This mirrors early hope for the liberatory functions of the Internet, in particular Stewart Brand. See Foer 2017: 1–32.

30. 'Surplus of sensibility' names the 'recorded behavioral traces' consistently harvested by our various digital prostheses (Hansen 2015a: 66). The surplus is fed back into predictive loops. It is both lived, insofar as it tracks the choices, and in the cases of physiological monitoring, the life processes of living subjects but it exceeds the intentional consciousness. When a digital device measures heart rate, for example, it produces a new form of sensibility even as this sensibility is founded on the life process of the individual.

31. Guattari's interest in the production of capitalist subjectivity bears relation to 'agnotology' as the study of the production of ignorance. However, where these analyses are epistemic in focusing on how ignorance is produced by a confluence of power and industry, Guattari is more interested in affective dimensions of the production, shaping and controlling of *desire*. See Proctor and Schiebinger 2008 and Sullivan and Tuana 2007 for two collections in this field.

32. 'Lifestyle anarchism' is preoccupied with the 'ego and its uniqueness' and for this reason affirms 'polymorphous concepts of resistance [that are] . . . steadily eroding the socialistic character of the libertarian tradition' (Bookchin 1997: 164). Bookchin implicitly aligns 'lifestyle anarchism' with Deleuze and Guattari: 'Where social anarchism called upon people to rise in revolution and seek the reconstruction of society, the irate petty bourgeois who populate the subcultural world of lifestyle anarchism call for episodic rebellion and the satisfaction of their "desiring machines", to use the phraseology of Deleuze and Guattari' (1997: 165). Bookchin is correct to recognise that an individualistic libertarianism is insufficient for transformation (as Massumi puts it 'Anarcho-libertarianism is anarcho-capitalism' (2018: 105)). However, while his characterisation may apply to some nominal Deleuze-inspired readers, he clearly has not considered 'desiring machines' in any detail. Appeals to the 'ego' as sufficient locus of liberation are emphatically not what Deleuze or Guattari are talking about.

33. If we simply understand dissensus as a rejection of perceived norms, this cannot create connections necessary for forging sustainable transformation. This is what Bookchin is worried about in his critique of 'lifestyle anarchism' and his contrasting call for a 'social anarchism' (1997: 164–8).

34. The Michelangelo work to which Whitehead refers is more commonly rendered in English: *Day* and *Night, Dusk* and *Dawn*

35. Deleuze's notion of the work of art as a 'bloc of sensations' that is a 'compound of percepts and affects' (WIP: 164) is resonant here. Like Whitehead, such a 'bloc of sensations' has the possibility of creating eternal moments: 'even if the material lasts for only a few seconds it will give sensation the power to exist and be preserved in itself *in the eternity that coexists with this short duration*' (WIP: 166). Indeed, for Deleuze, this is the goal of art: 'to create the finite that restores the infinite' (WIP: 197).

36. 'Sedentary' distribution distributes an already occupied space, divvying it up according to 'fixed and proportional determinations'. By contrast, 'nomadic' distribution refers to a filling of space that is not already possessed. This is the difference between being 'distributed in space' (nomadic) and 'distribut[ing] the space' (sedentary) (DR: 36).

37. Stengers's primary concern is scientific practices, but she recognises that 'the constraints found in requirements and obligations do not in themselves singularize scientific practices any more than they do the practices [referred to] as modern' (2010: 54). Provided that we understand them with the appropriate measure of abstract flexibility, we can use obligation and requirement as orienting operations in looking at a variety of relations and practices.

38. While obligation and requirement are correlative concepts, this does not mean they are symmetrical (Stengers 2010: 52). Certain events enact large obligations with minimal requirements. Stengers also distinguishes obligation and requirement from 'rights' and 'duties': 'requirements and obligations do not function in terms of reciprocity and, as constraints, what they help keep together is not a city of honest men and women but

a heterogeneous collective of . . . phenomena whose interpretation is at stake' (2010: 52).

39. Practical, epistemic and existential do not demarcate exclusive categories, since any event could manifest aspects of all three.

40. Stengers is quoting Guattari's *Chaosmosis* (1995) here.

41. Note that such an analogy risks overlooking the degree to which an achievement or event is frequently between what would be commonly thought of as different subjects.

42. See ATP: 502–3; Stengers 2010: 266, n11. This reference can be more or less precise and is not symmetrical. In the case of a predator/prey relation, for example, the prey may employ strategies of evasion specifically tuned to the cognitive capacities of its standard predator, while the predator may have a range of prey options that it does not need to meaningfully disambiguate. Nevertheless, its identity as predator is always in relation to a prey.

43. Stengers refers to a species of bat that does not appear to play any significant role in the ordinary functioning of the Puerto Rican tropical forest that is its home. However, the species plays a highly significant role when the forest is devastated by a hurricane, since it is incapable of flying away and relocating. The ecological function of the bat is negligible on an ordinary temporal scale but appears during periods of crisis.

Chapter 6

The Risks of Affect

The double bind of ecological attunement

From different routes, I have repeatedly emphasised attention's ontological force. The results of choices of attention have effects in what Whitehead calls the 'passage of nature' (Chapter 4). Through emphasising what remains its speculative potential, it has been necessary to detach attention (as concept) from its ordinary contexts. Attention becomes ontological when understood as participant in the 'concrescence' of events and occasions whose concrete actuality is prior to the abstractions of Self-I-Subject. Attention is thus an existential mode through which it may be possible to become closer to living reality's individuations. Deleuze writes 'it is the I and the Self which are the *abstract* universals' (DR: 258). However, this does not mean that undoing this abstraction enables us to find a truer, more primary and concrete self beyond these abstractions. Rather, 'Beyond the self and the I we find not the impersonal but the individual and its factors, individuation and its fields, individuality and its pre-individual singularities' (DR: 258).

Can we cultivate capacities of attention to approach individual intensities operative beneath ordinary forms of convention and intelligibility? If it is true that this potential is most pregnant in an intensification of sensibility's encounter with feeling or tertiary qualities prior to conceptual representation, such that feeling and tertiary qualities become indicative of more than simply 'subjective' taste, it is also the case that feeling or the tertiary quality do not automatically manifest in this way (Chapter 3). Understanding feeling, affect and their expressions in tertiary qualities as ontologically active does not automatically translate from their ordinary apprehension in emotions.[1]

Finally, as we have seen (Chapters 4 and 5), attention's events do not develop or occur in a neutral or 'pure' experiential field. Rather, attention's ontological potentials are threaded through powerful procedures of control manifest in the way a logic of equivalence constrains processes of psychic ecology. Such logic diminishes rather than intensifies the stakes of attention, encouraging closure to the potentials of encounter with difference or surprise. Through equivalence, it is always more of the same. Coupled with statistical summary and algorithmic data-mining as these feed into loops of predictive technologies and digital prostheses, the challenges in fostering existential capacities for unleashing creative attention are significant.

These challenges are affective, social, epistemological, axiological and existential. They are affective, insofar as equivalence conflates security with predictability. Opening to attention's creative potentials is necessarily risky, since it may unsettle perceived stabilities and identities (Chapters 1 and 2). They are social, since the economic and material operations of Integrated World Capitalism are thoroughly suffused with procedures of equivalence. These social challenges result from the conforming pressures of common sense, especially as intensified through the echo chamber phenomena of digitalised social media. To attend differently is to risk incoherence, exclusion or the presumed capture of a polarised social field, where if you are not 'like us', you must be 'one of them' (always the presumption that reality is finished, determined, and that all phenomena, subjects, objects, fit into a pre-existing rubric of order). They are epistemological, since they require relinquishing the presumption that knowledge is best defined through its certainty (Chapter 1). They are axiological, since creative attention requires openness to the emergence of unexpected values or meanings that can only retroactively be ascertained. Finally, there are the simple yet profound challenges of fatigue and finitude.[2] There is always more to be attended to: we can never achieve once and for all a maximal flowering of creative attention that I have christened ecological attunement. But its potential, as ideal or 'propositional lure', is always there, beckoning, even as the achievement of any single moment of attention is given its definiteness through gradations of exclusion (Whitehead's 'negative prehensions').

Ecologically, a salient challenge is in spatial-temporal scale since a processual conception of individuation means there is no unique serial order of time, but rather many different durational times. The 'living present' differs in relation to perspectives: the mountain, the redwood, the hummingbird, the infant, the elder, the electron live different presents. And yet, this difference is not exclusive, since these differing presents converge in each actual occasion. The *differing* of perspectives is thus not a function of their absolute separation; rather, it is precisely this differing which makes each of them an ingredient in concrescence as a creative addition. That is, if the presents did not differ, the creative addition would achieve less intensity. And yet, this remains difficult to construe existentially, especially in relation to temporality, since time, as time, is often taken as lacking any quality in-itself. Events occurs in time, and they may be frightening, joyful, and so on, but time remains the pure and empty form running beneath these events. To acknowledge that this view of time is rejected by both Deleuze and Whitehead is not yet to make its existential implications manifest.

Can we make sense existentially of the idea that our actions ripple across multiple scales of space and that different temporal durations intersect in a living event? Ruyer suggests that because 'the core of the theory of (special) relativity suggests that we cannot be in two locations at once . . . the absolute subjective expanse escapes the theory of relativity' (2016: 94). But I am suggesting a different strategy, following Whitehead and Deleuze. The absolute subjective expanse, if there is such a thing, is not equivalent to ordinary lived experience. Indeed, such experience is almost entirely mediated by abstractions that in effect hinder or obscure existential encounter and participation in the occasions constituting that experience. The notion that we do not live in a single order of linear time, but in many different times and tempos that need not have a single common metric unifying them ('being in two or more times at the same place'), is a conceptual call to wake up to the qualitative complexity of the present.[3] Attuning to this complexity means opening to a precise singularity that is uniquely situated and then gone. But the quality of this singularity is an expression of encounters that resonate beyond the here-now. Negotiating this means being mindful of differently experienced pasts, differently thickened presents, and differently projected futures conspiring together for each achievement of actual reality. There is no master perspective that can univocally narrate what 'really happened' – which is not to say that it is all made up, but rather that events continue to take a long time to happen.

These challenges intersect in an existential complex for which there is not adequate language: pragmatic-spiritual-ethico-aesthetic? Such a problematic complex is real. If one's image of thought is oriented towards immunisation from the problems of living, the consequences of the view developed here are untenable. Rather than getting rid of the problems, this makes them more palpable! In opening the abstraction of a stable Self to its impersonal individuations, the double bind of an ecological attunement is heightened. It is not that the Self, as is, is redemptively incorporated into a larger harmonious Whole. Quite the contrary, if from one perspective the Self is a 'fiction', it is still a living idea to which powerful attractions accumulate and is in this sense quite real. Moreover, attending to the collaborative relationality of emerging actual occasions means accepting one's lack of power in fully controlling their outcomes while heightening responsibility by altering the stakes of attention from passive receptions of a world already determined and made up in advance to active collaborations that feed forward in partially binding ways for future collaborations. Less control, more responsibility! Who wants that?

Such constraints ensure that ecological attunement cannot result in a definitive normative prescription in the form of static categorical judgements. There are no merely neutral attendings in the ontological sense, but there is also not one right or 'correct' way to attend or create. What would this even mean? But giving up static categorical judgement is more easily done at the level of theory, in living practices one still has to choose and act. The risk for an abstract conception of ecological attunement in resisting definitive normative prescriptions is the failure of viability in inspiring the work of transforming attention. How does a concept of ecological attunement negotiate such mutual constraints: the necessity of choice and action with the rejection of universal categorical prescriptions for action?

Attending to tertiary qualities heightens attunement to relationality beyond identity boundaries when we understand those qualities to exceed the merely subjective. But this means we have to foster a sensibility that is always alert to its own tendencies to reduce its encounter with the real by presuming a neatness of fit with its pre-existent categories. This aspirational conception of the potential for such sensibility is a clear point of resonance between Whitehead and Deleuze. But this appeal to an ontological conception of feeling/affect, expressed as I have in the language of tertiary qualities, has significant risks as well. Such risks must be lived; that is, any attempt to deny them is already to use thought to shut down the 'plane of immanence'. (This is another way of saying that the risks are metaphysically real, not simply the result of errors following human limitation.) But conceptual work in more precisely articulating some of their contours can help this living negotiation be more aware of its blind spots or pitfalls.

Attunement and affect: vitality *before* world

Ecological attunement is a manner of attention that follows from a processual metaphysics. Disambiguating a philosophical concept of ecology from its scientific function meant incorporating psychic, social and material processes into its purview and hence brought us to Guattari's 'existential territory' (Chapter 5). This concept can seem primarily phenomenological, merely a new phraseology for the life-world (Husserl's *lebenswelt* or

Heidegger's *Dasein* (there-Being)). This appears even more plausible given the use of 'attunement', a term whose most famous deployment in philosophy is certainly Heidegger's.[4]

While fully working out the differentiation with the complex phenomenological tradition is a matter for another book, there are several important ways in which 'attuning' here functions differently than *Die Stimmung* or mood (in Heidegger primarily). Articulating these differences, while not intended as an exhaustive engagement with Heidegger or existential phenomenology, is nevertheless helpful for moving the speculative proposal I am developing. Most notably, mood or attunement (*die stimmung*) is a primary concept Heidegger uses to distinguish between Dasein's being in a 'world' and what he calls the nonhuman animal's 'poverty of world'.[5] Mood or attunement is not 'one existing property that appears amongst others' (Heidegger 1995: 65); rather, attunement is linked to the very being of Dasein (there-Being). In *Being and Time*, Heidegger declares mood prior to representative cognition as a 'fundamental existentiale' that offers a 'primordial disclosure . . . in which Dasein is brought before its Being as "there"' (2008: 134). He contrasts the ontic or everyday sense of having a mood (today I feel happy, depressed, and so on) with its ontological condition. While we may, 'factically' seek to control or modify our ontic mood, this does not deny that, ontologically, attunement or mood is prior to conceptual control. Mood thus 'discloses Dasein in its thrownness' (2008: 175/136). Mood is like the primary filter or sieve through which everything that shows up is mediated or tinted, like an effect on a lens.

As a fundamental existentiale, attunement is always linked to world for Heidegger. Indeed, it is through attunement that Dasein opens up world. In *The Fundamental Concepts of Metaphysics*, this opening is in contrast to the nonhuman animal. The nonhuman animal does not attune, it is rather captured by a limited number of viable 'signs' that correspond to its drives and instincts. Heidegger characterises the condition of the animal as 'captivation': 'the animal can only behave insofar as it is essentially captivated' (Heidegger 1995: 239). Because animal behaviour is captivated according to its drives, it does not open a world: 'the behavior of the animal is not a doing and acting . . . but a driven performing' (1995: 237). This is why Heidegger declares that 'the animal behaves within an environment but never within a world' (1995: 239). For world, a being must have the capacity to 'apprehend something as something, something as a being' (1995: 264).

In this sense, Heidegger's attunement remains firmly in the tradition of humanism. Even as he offers a novel distinction between human and nonhuman that does not rely on rationality or language, he is still committed to their fundamental ontological difference. By contrast, ecological attunement, which in its strictest terminology is always ecological attun*ing*, shifts the locus from Being to events and occasions. This has important consequences. Attuning keeps the musical connotation of *Die Stimmung*, but expands the players, if you will. Ecological attuning is thus ontologically prior to the constitution of a 'world'. Deleuze and Guattari's discussion of the territorialising operation of the refrain, which is an ontological discussion, makes this clear.

Deleuze and Guattari link the refrain's territorialising to what they call the 'becoming-expressive of rhythm and melody' with a capacious and cross-modal understanding of rhythm to include 'the emergence of proper qualities (color, odor, sound, silhouette)' (ATP: 316). This becoming-expressive involves a musical collaboration in which a territory is constructed on the basis of 'expressive qualities'. In contrast to a fundamental attunement which hovers like an affective sieve across the reception of all qualities, attuning to expressive qualities can never be characterised through one fundamental affect. Or, if it is characterised by one fundamental predominating affect, this is an indication of a diminishing refrain, a society tending towards death, as Whitehead would say. World, if we insist on that term, is always commencing, because it is always in relational contact with more-than-world, with outside-world, with other-world, and so on. For Deleuze and Guattari, who use 'territory' rather than world, the becoming or emergence of proper qualities (what I have referred to as tertiary qualities) can be called 'Art': 'the territory [is] a result of art' (ATP: 316). Moreover, a territory is neither finished nor possessed: 'the expressive is primary in relation to the possessive; expressive qualities, or matters of expression, are necessarily appropriative and constitute a having more profound than being' (ATP: 316).

When Deleuze and Guattari say that these expressive qualities are 'necessarily appropriative', they immediately insist that this is 'not in the sense [of] belonging to a subject, but in the sense that they delineate a territory that *will belong* to the subject' (ATP: 316, emphasis added). These qualities are 'signatures, but the signature, the proper name, is not the constituted mark of a subject, but the constituting mark of a domain, an abode' (ATP: 316). This is to say that one does not project identifying qualities (like a primordial mood) in front of them like a cloud of possession, rather, the qualities come first, and in the activity of attuning a domain is created in relation and dialogue through these expressive qualities. This is why, for example, 'expressive qualities . . . enter shifting relations with one another that "express" the relation of the territory *they draw* to the interior milieu of impulses and exterior milieu of circumstances' (ATP: 317, emphasis added). Ecological attuning is thus a manner of attending to the edges between territories and milieus. The expressive qualities (what I have called tertiary qualities) function as 'signs' that cross milieus and are thus the best access we have to the ontological constitution of the occasion in its creative sense. This is why they represent a 'having more profound than being' – it is a having in the sense of a constitutive experience that has (creates) one, rather than that is. Having is dynamic, being is static.

This expressive interplay of refrains and qualities ('expressive qualities entertain variable or constant relations with one another' (ATP: 318)) is inherently ecological in undercutting the primacy of any single relation and opening to a dynamism of multiple relations with various degrees of stability. Rather than constituted subjects and objects, 'we no longer have the simple situation of a rhythm associated with a character, subject, or impulse'; instead, 'the rhythm *is* the character' (ATP: 318). Ecological attuning expresses a manner of attending to these rhythms (tertiary qualities) as primary ontological modes. This entails a gestalt shift or inversion of foreground and background. Sounds, colours, affects, shapes, tones, smells are no longer the accidental or contingent, they are what the landscape or ecology does: 'the melodic landscape is no longer a melody associated with a landscape, the melody itself is a sonorous landscape' that is 'in counterpoint to a virtual landscape' (ATP: 318). Always, it is a question of counterpoint, of changing relations, of the way in which an expression ripples through in series of occasions and events, which set forth further series and occasions. The constants are not the objects and subjects, they are the tonal affects, which function like potentials that actualise according to different intensities:

> We can say that the musician bird goes from sadness to joy or that it greets the rising sun or endangers itself in order to sing. None of these formulations carry the slightest risk of anthropomorphism . . . It is instead a kind of geomorphism. The relation to joy and sadness, the sun, danger, perfection, is given in the motif and counterpoint, even if the term of each of these relations is not given. In the motif and the counterpoint, the sun, joy or sadness, danger, become sonorous, rhythmic, or melodic. (ATP: 318–19)

Ecological attuning understands its own attendings as participant in this counterpoint. One's affective and material affordances are of course constrained – there is no maximal bandwidth that somehow envelops the whole, at least for the human being. But such bandwidths can make more or less conscious music, and this making is, in some partial but never fully determinate sense, a function of attending – the 'to-what' one is attuning. The aspirational goal, in the ecological sense, is to understand this attuning as ultimately linked to the 'forces of air and water, bird and fish' such that one becomes an expressing modality of the 'forces of the earth' (ATP: 321). This is what Deleuze and Guattari mean by 'becoming-imperceptible'.

If 'becoming-imperceptible' presents a lure for living, can we articulate more concretely how this might affect living habits? Or, alternatively, is there a way of drawing out what a notion of attuning to an ecology of expressive qualities could mean in other domains of theory or thought?

Cynthia Willett draws on a non-Heideggerian conception of attunement to conceive of forms of communication across species boundaries that she describes as 'proto-conversation' (2014: 82). If attuning is less a primordial orientation and more a relational modality *between the expressive qualities of juxtaposed moments*, then it is also potentially operative across boundaries between different forms of life. Note that this possibility becomes even more cogent under a process metaphysics, since the forms of life are not discrete and separate entities, but rather routes of historical inheritance (societies) defined, in a sense, through their reception and transformation of previous occasions. This is not just to affirm nonhuman animals as expressing their own modes of meaning-making, play and intersubjectivity, but to raise the possibility that such modes might be collaborative with an ultimate interest in exploring an ethics based on a 'biosocial eros' (Willett 2014: 82).[6]

Ecological attunement cannot follow Willett's emphasis on the ethical as yet. While this is a noble goal with which I am in much sympathy, I do not find an unequivocal ethical implication *at the level of theory* to follow from ecological attuning.[7] That said, Willett's project, and especially her use of Daniel Stern's work studying 'affect attunement' between human infants and adult caregivers, offer interesting examples of how an ecology of expressive qualities and corresponding processes of subjectification as refrains in counterpoint may illuminate more empirical work.

Stern articulates 'affect attunement' to express how affect patterns transfer without being threaded through the conscious intentionality of a subject (since the newborn infant does not yet developmentally have a sense of self). His later work draws more extensively on cognitive science and expands the concept to include attunement to what he calls 'vitality forms'. Three features of Stern's account are important for ecological attuning. The first is that attunement is cross-modal in that it can occur across different perceptive modalities, transferring for example between sound and sight (Stern 1985: 154–7, 2010: 26–8).[8] The second is that attunement is distinct from empathy, mimicry or mimesis. Though affect attunement 'starts with an emotional resonance', it 'does something different with it . . . [it] takes the experience of emotional resonance and recasts that experience into another form . . . that need not proceed towards empathic knowledge or response' (Stern 1985: 145). This distinction is one of the places where Stern's concept is better enabled by a process meta-physics rather than a more traditional substance approach. With the latter, empathy presumes a sharing of predicates between stable subjects: you feel angry, and now I feel angry. The emphasis is on the identification of two shared emotional states that function as the same predicate. With attunement, however, the emphasis is on a manner that passes through both but does not presume an equivalent affective identification. Attunement as a mode of transfer thus allows for shared experiences to constitute different subjects that emerge as altered through their relation.[9] Rather than identification, 'attunement is a distinct form of affective transaction in its own right' (Stern 1985: 145).

An affective transaction which does not presume a logic of identity is attractive for ecological attuning, but also emphasises its normative challenges.[10] Attunement does not function as a means of consolidating meaning that is the same across difference, but as a way of connecting or bringing into relation. But the meaning of such relations is not necessarily shared or agreed upon.[11] This brings us to the third important point, which is that affect attunement is not necessarily intentional or conscious. Stern writes: 'evidence indicates that attunements occur largely out of awareness and almost automatically' (1985: 145, 152–61). For the argument of this book, which is exploring the way in which a different meta-physical understanding might shift habits of attention, the 'almost' in this sense is important. If attunement was simply and always automatic, then there would be no point in developing a concept of ecological attuning. That said, it is clear that attuning is not primarily an activity of the conscious will. For the gambit of this text, it requires shifting one's habits of attention and being more open to differences and gaps where one does not know, but one endeavours to feel or encounter, expressive qualities without automatically enlisting them into one's prevailing narratives.

Stern's disambiguating between emotions and 'vitality forms' is important for more speculative applications of attuning like Willett's since it involves opening to relations across differences where ordinary concepts of emotion do not apply. The object of attunement is not a discrete emotional state, but rather a style of expression actualised through rhythm, timing and intensity. In his earlier work, Stern distinguished between 'discrete categorical affects' and 'vitality affects', later to be renamed 'vitality forms'. Discrete categorical affects refer to experiential states *commonly understood as emotion*: happiness, sadness, fear, anger, disgust, surprise, interest, and so on, and it is these that would be relevant for concepts of empathy and mimesis.[12] Vitality affects however are qualitative forms: surging, fading away, crescendo, bursting, etc. (1985: 54–5, 2010: 23–8). A vitality form is a dynamic pattern that expresses shifts without *predetermining the emotional content of these shifts.* While this is not an exclusive distinction ('vitality affects occur both in the presence of and in the absence of categorical affects' (1985: 55)), attunement functions at the level of vitality affects, not categorical affects. Stern's later work explores this point in greater depth, studying how 'vitality forms' inform a wide range of experience. Stern insists on the distinction from emotion: 'vitality forms are different from emotions in their nature, feel, non-specificity, omnipresence and neurobiology' (Stern 2010: 28).

Willett is interested in 'vitality affects' or 'vitality forms' in providing a way of thinking relational encounter not mediated through a shared 'world' in the phenomenological sense. Vitality forms are not equivalent to emotional content. Stern tends to discuss them in terms of manners of intensity, timing and rhythm. These forms need not be states of a presumed subject; but attuning to them brings one into greater contact with the world around them. The maple tree's surging burst into autumn red becomes felt in a visceral way. This is not a projection of one's emotional state but a receptivity to an expressive quality. Furthermore, combined with a processual metaphysics, such attuning might be one way of gaining greater awareness, if not control, of the creative effects of attention – as one becomes more conscious of the feedbacks of attention – how one's choice to attend, or not attend, to expressive qualities influence subsequent attuning and the created quality of existential territory.

Understanding vitality forms in such a metaphysics complicates the Kantian-Uexküll transcendental pluralisms in which each species, agent or subject only perceives the forms available to it.[13] If vitality forms partially characterise transfers from occasion to occasion they are prior to the distinctions between organisms even as that eventual distinction emerges out of these transfers. Finally, and most speculatively, vitality forms give a way of thinking attuning that is prior to 'world' and as such open one to encounters beyond the preconceived. Such a capacity may be all the more important in periods of duress and destabilisation, since they enable one to attune to what is vital in a situation even as its ordinary markers are dissolving or changing. However, and this is the promise and peril of vitality forms, because attunement is not always intentional, there is an inherent risk for manipulation and capture, as we see in the charisma of fascist leaders for example. So, ecological attuning must explore its normative complexity in more detail, even if the outcome is not a categorical judgement.

Tertiary qualities and Self/Other

Can ecological attuning, combined with a speculative notion of vitality forms or asubjective tertiary qualities, make sense of Guattari's call for individuals to become both more united and increasingly different? How can we understand such a call without presuming a falsely harmonious holism? As Whitehead reminds us, 'a characteristic of every living society is that it requires food . . . [this requires] interplay with their environment . . . [that] takes the form of robbery' (PR: 105). Is there a way of conceptualising encounters with difference without recoiling into antagonism while remaining cognisant of 'robbery'?

Ecological attuning disrupts homogenising habits of attention captured by equivalence. In the context of a processual concept of subjectivity, the stakes of this disruption are ontological. In attuning, we do or do not develop capacities and affordances that become both enabling and constraining for manners of attention going forward. However, if there is *a degree* of creation in each occasion, this degree is often heavily constrained. Moreover, such creation is not necessarily aligned with our perceived interests, whether personal or social. Citing Whitehead's claim that 'God's purpose in the creative advance is the evocation of intensities', Stengers observes that this means that 'God is indifferent to what counts for most of us: preservation' (2011: 317).

Such evocation of intensities is in tandem with the dissolution of a certain concept of subjectivity, as explored in the first part of this text. Subjectivity has been redescribed as a manner of consistency exhibited in patterns of relational feeling – such that it is not that we 'have' feelings, but rather that feelings have us. But how does this conception manifest in more ordinary modes of speaking, thinking and being in relation to others? It is one thing to understand the speculative posit of the actual occasion as related to all other actual occasions in gradations of relevance, but what does this mean for the person currently struggling to understand their interlocutor?

Deleuze's brief development of what he calls the *a priori* Other structure expresses the extent to which a processual conception of subjectivity departs from entrenched scripts while also raising potential problems. Of most significance is how his characterisation follows from an affirmative, rather than negating, conception of difference. At its best, as Rosi Braidotti observes, this offers a thinking of subjectivity in a manner alternative to the Hegelian influence in critical thought, which underwrites a theorising of the relation between subject and other through a logic of negation: you are who you are because you are not that other. Not that, not that, not that – not them, not them, not them. By contrast, a process conception of subjectivity resonates with what Braidotti calls a 'nomadic subject' that 'disengage[s] the emergence of the subject from the logic of negation' (Braidotti 2011: 323). At its worst, however, Deleuze's way of describing the *a priori* other structure may reinforce the risk of a solipsistic self-referential affect, since it appears to inscribe the other as primarily a structure of possibility for the Self.[14]

In *Difference and Repetition* and an appendix to *Logic of Sense*, Deleuze articulates the *a priori* Other structure as a 'structure of the possible' (LS: 307).[15] This structure is a transcendental condition of the development of the psychic system commonly called self, precisely because 'the self is the development and the explication of what is possible' (LS: 307). It is because the Other structurally expresses the different that it is a necessary condition for the development of the psychic system of Self. Without such difference, there is no disruption to the Self's perceived world and thus no differentiation: 'the Other is initially a structure of the perceptual field, without which the entire field could not structure as it does' (LS: 307).

The Other does not function as an Object or a competing Subject, but rather as an expression of a 'possible world' that is characterised affectively through tertiary qualities. Deleuze gives the example of an encounter with a terrified face when the perceiver does not experience the cause of terror: 'the face expresses a possible world: the terrifying world' (DR: 260). This expressive potentiality gets explicated or unfolded through the encounter as a mode of individuating the self: 'the terrified face does not resemble what terrifies it, it envelops a state of the terrifying world' (DR: 260). The Other is not simply transmitting an external condition for the unfolding Self to see on their own, rather, the *a priori* Other structure is defined through its manner of concretising difference. This difference is both in terms of the encounter between Self and Other and in the expression of Other to the world: 'The terrified countenance bears no resemblance to the terrifying thing. It implicates it, it envelops it as something else, in a kind of torsion which situates what is expressed in the expressing' (LS: 307).

If we read Deleuze presuming the Self exists independently prior to this encounter, then his description is open to a criticism whereby the Other is simply the means by which a Self achieves autonomy. Alice Jardine presents a strong criticism along these lines from a feminist perspective, also considering how 'becoming-woman' in *A Thousand Plateaus* similarly perpetuates a masculinist logic where the route to an authentic individuation proceeds through the Woman.[16] Such worries are important, but this is not the only way to read the *a priori* Other structure in the context of a speculative metaphysics where the autonomy of Self is an illusion if taken as metaphysically complete. Here it is important to maintain a distinction between an individuating and the psychic system that becomes characterised and identified as Self. The former is metaphysically primary and not equivalent to the later.

With this distinction in mind, what is primary in the encounter expressed in the *a priori* Other structure is the tertiary quality – not the Self or the Other. As Deleuze says, 'Concretely, it is the so-called tertiary qualities whose mode of existence is in the first instance enveloped by the other' (DR: 260). Otherness is not given or located at the level of persons, but rather as expressions of differing values: 'the a priori Other is defined in each system by its expressive value – in other words, its implicit and enveloping value' (DR: 260). While it is true that the Self often operates with a falsely reified sense of this constituting system, Deleuze is interested in the extent to which the system includes a necessary disruption to autonomy through the encounter with Other such that both Self and Other are expressions of implicating values and intensities. What is really Other is not the person, so much as the tertiary quality. This otherness disrupts the stability of both sides of the encounter: Self and Other.

What is ontologically prior are the tertiary qualities as mode of encounter between: 'Fear', 'Terror', 'Joy', 'Curiousity' are actualisations of differing intensities around which the self/other structure gets explicated:

> The I and the Self, by contrast, are immediately characterized by functions of development or explication . . . they tend to explicate or develop the world expressed by the other, either in order to participate in it or to deny it. (DR: 260)

Jardine's worry however remains, precisely because these metaphysics push on ordinary lived conceptions of subjectivity to such an extent. Deleuze indeed recognises this when he acknowledges that the I and the Self either 'participate or deny' the 'world' of the Other. Deleuze's account can only escape these worries if it is understood to function at the level of events, not the level of ordinarily lived subjectivities. Deleuze however says as much: 'In order to grasp the other as such, we [have to] insist upon special conditions of experience . . . the moment at which the expressed has (for us) no existence apart from that which expresses it: the Other as *the expression of a possible world*' (DR: 261). The *a priori* other structure offers a clue (a sign) for other (different) ways to affirm the event's excess to any single perspective on it. In this way, the encounter with the Other may function as a 'dark precursor' or 'quasi-cause' (Chapter 2) in triggering a threshold and as such changing the trajectory of the 'I-Self' system. The other as such is expression of a quality, not the world, and not the person. This is why Deleuze says that 'In the psychic system of the I-Self, the Other functions as a centre of enwinding, envelopment or implication. *It is the representative of the individuating factors*' (DR: 261). These individuating factors express in experience as tertiary qualities or asubjective forms of vitality prior to their capture in consolidated identities. When we experience the encounter with an otherness of feeling, the structure disrupts the Self's homogeneity and (potentially) pushes it to unfold more intense or differing values and 'grow' or change. Without this encounter with alternative possible worlds, the world becomes static, inert and equivalent to one's interiority: 'The Other assures the margins and transitions in the world' (LS: 305).

This does not mean that such an encounter is necessarily harmonious nor always 'positive'. Nor is such an encounter easily assimilable into norms of understanding or transparency. What Deleuze is interested in is the extent to which such an *a priori* Other structure challenges the idea that the encounter results in a shared consensus or explication as to its meaning. This is why he warns against over-*explication*, both with regard to Self and Other (DR: 261).[17] Because it functions at the level of the extensive and constituted forms of representation, explication has a tendency to 'cancel' the differences of intensity that are its condition of possibility: 'The hard law of explication is that what is explicated *is explicated once and for all*' (DR: 244). Explicating the other strips their power to produce difference. It does the same to the self. Always something must be 'kept in reserve' (DR: 244). Indeed, Deleuze characterises this as an 'ethics': 'affirm even the lowest, [and] do not explicate oneself [too much]' (DR: 244).

Such affirmation can seem a wilful obscurantism if one does not keep in mind the metaphysics in which it arrives. Explication is a mode of separating, whereas for Deleuze the question is always how to construct greater (more intense) modes of resonant connection. The operation of such connection is through sensibility, not intelligibility. The question for an ecological attuning then is whether or not we can see the *a priori* Other structure as operative in a wider field than just our immediate acquaintances and species.

But there is a second, and perhaps more important, *temporal implication* that follows this characterisation of the *a priori* Other structure. Indeed, because of the risks of encounter, because we cannot presume a level playing field, and because, as Whitehead notes, we both eat and are eaten, the normative question cannot be jettisoned. Even as we affirm the encounter and the *a priori* Other structure as emphasising lived importance of the tertiary quality as a sign of individuating factors, we also live these encounters in a dangerous world. The Other expresses a possible world that can kill, eat or otherwise injure you. Moreover, while from a metaphysical perspective we live on the edge of these encounters, they are also continuously being incorporated into the molar lines of subjectivity's ordinary representations. What can this mean for ecological attuning? While both Self and Other must resist complete determination, representation or understanding, they are nevertheless also always subject to relative determinations, relative representations and degrees of understanding. But such relative determinations, representations and understanding manifest retroactively, when the occasion is past. As Deleuze puts it: 'If the Other is a possible world, I am a past world' (LS: 310). If descriptively this follows a processual account, it appears to strand the would-be attender in a place of passivity: where they can never know how the encounters in which they are immersed will turn out even as they must choose what to affirm in them.

Janus-faced time: selection and orientation in ecological attunement

Ecological attunement alters the stakes of habits of attention. Instead of passive reception of a world determined in advance, attention is implicated in active collaborations that feed forward in partially binding ways for future collaborations. However, while there are no merely neutral attendings, there are also no general prescriptions to be extracted as static universal principles. In some sense, the occasion itself engenders its own value. What can this mean in a practical sense? Life requires choices and actions, many of them fraught with uncertainty. How does a concept of ecological attunement navigate this double bind: the necessity of choice coupled with the impossibility of universal categorical principles?

If a 'Self' is an abstraction from relational patterns exemplified in tertiary qualities, then attending to such qualities becomes a way to 'know thy-self' and connect this knowing to larger ecological contexts. Tertiary qualities are by definition relational. Formally, Locke's characterisation tends towards the two term or dyadic (meltable/melt-causing), but this cannot hold in a strict sense. In a process context, even tertiary qualities are better understood as manners of qualifying events. They thus name a relational pattern that repeats but can be realised in different intensities. This realising is constitutive of the occasion and functions at the intersection of the axiological and the aesthetic. The explosive tension of a fraught encounter, pervasive throughout the room, is a form through which disparate series constituting the event of the room connect. Feeling is shared, but how the feeling manifests is not.

The boundaries of this shared activity of feeling do not map the boundaries of the organisms. Moreover, within a Whiteheadean framework, the subject/object pivot is no longer correlative to the Self/Other, but rather is incorporated into the structure of actual occasions and the grain of experience. Explosiveness is an inheritance of the occasion, and as the occasion exemplifies this quality, along with others, it is its own subject. In its completion, the occasion becomes objective for further occasions to inherit. Each actual occasion is shaped by the relations it inherits, and this inheritance is both informed by, and the result of, the tertiary qualities mediating between occasions. The feeling is not an add-on to an occasion that happens, but rather a mode of actualising the occasion's unification. For this reason, the more capacious a range of feeling is, the more complex and intense the occasion can be, because it operates by achieving a moment of unification across a greater range of differences that produces a qualitatively more complex intensity.

Are such moments always 'good'? This is not a well-formulated question since it supposes the possibility of a monolithic judgement. Temporal heterogeneity (good when and at what register or scale?) and dynamism (good going in what direction?) problematise such a single-register question. However, in attempting to articulate practical dimensions of an ecological attunement, there is still some vestige of existential meaning. From this vantage, even while admitting the reduction induced by asking the question, the answer is clear. The relational structure of moments out of which a life is knit means there is no conclusive way to immunise from conflicts and discords. In short, no, such moments are not 'always good', harmonious or constructive.

The question is poorly formulated from a metaphysical perspective because it presumes the possibility of isolating a dominant stable perspective as ground from which to judge. But given the heterogeneity of scales involved in processes of individuation, such an isolation will be a reduction. What is one to do? And the paradox gets more intense, precisely because the *achievement of a genuine perspective is in some sense the goal of ecological attuning.* Everything depends on what is meant by genuine. That is, a perspective can simply conform its inheritance and follow the path of least resistance in being assimilated into trends of equivalence, in which case the opportunity for the universe to express itself more capaciously is diminished, or, it can understand its own processes of attention as forms by which the universe is singularised. How does one include as much as possible without disintegrating? And how do we understand inclusion, in a singular moment of individuation, while also keeping in mind the multiscalar complexity of ecology?

The most pertinent lived dimension of this complexity is temporal. For this reason, ecological attuning requires dual modalities, selection and orientation situated along two different temporal axes – that is, one looks backwards and one looks forward.[18] It is true that attuning as activity is always enacted in the 'living present' that Deleuze refers to as the first synthesis of time (DR: 70). The living present is constituted by the passive synthesis of habit. Importantly, the 'subjects' of these habits are not constituted selves but rather 'the primary habits that we are; the thousands of passive syntheses of which we are organically composed' (DR: 74). Deleuze calls these passive syntheses, which in many respects are akin to Whitehead's actual occasions, as 'larval subjects' or 'little selves': 'Underneath the self which acts are little selves which contemplate and which render possible both the action and the active subject' (DR: 75). In insisting on the ontological dimensions of attention, attuning is made up of the corresponding micro-contemplations of these passive habits. However, insofar as we seek to alter their enactment in the present, we must insist on a possibility of learning how to alter or transform their activity. This cannot be an act of direct conscious will. But it may be possible, through reflection, to influence these constituting habits. Indeed, when Deleuze observes that 'it is simultaneously through contraction that we are habits, but through contemplation that we contract' (DR: 74), it is this possibility that activity of contemplation can influence contractive habits that is opened.

And yet, this does not mean that this influence is direct or unequivocal. Everywhere there is uncertainty and ambiguity. We are creatures of contractive habits that form without our realising it. It is not just a matter of changing habits but of knowing what habits to change, and at the level of the 'little selves' it cannot be assumed that these contractions and contemplations mirror intentional normative valuation. The fundamental issue is in terms of the tendency of a habit – does it move towards an intensifying or diminishing of contemplation's awareness? Intensifying awareness must be taken as a good, even if the objects of awareness are not always harmonious. To be more aware is to be more alive is to be more potentially responsive and creative. But how is contemplation intensified or diminished by choices of attention? Can we learn to notice the 'feeling' of such intensification or diminishment?

Ecological attunement, as concept, requires a reflective looping that, it is hoped, can over time infiltrate the passive syntheses of the living present. Consciously, this involves observing the effects and outcomes of habits of attention and learning to better discern their influencing manner on the quality of one's present. This discernment must also remain a process, given the varying temporal feedback loops in question. This temporal complexity admits of two main modalities: a retroactive selection and a futural orientation.

Consider how the distinction between Chronos and Aion inflects Deleuze's discussion of an existential attitude in relation to the event.

Deleuze adapts this terminology from his discussion of Stoic ontology where 'time' functions as one of the four 'incorporeals' (void, place, time, sayable) contrasted with bodies or states of affairs (LS: 7–23).[19] For the Stoics, the latter are a complex, even cosmic, whole. Differentiations between different bodies can be made, but they require the intervention of 'incorporeals'. (This is why 'void' and 'place' are 'incorporeals' – because conceptually they require a place outside of bodies or states of affairs from which to locate or differentiate these states of affairs.) In bodies themselves, there is only the cosmic whole.[20] The 'incorporeals' thus are the means by which Stoics explain the thinkability of this whole from the limited vantage of the finite human. With regard to time, this entails a further distinction: Chronos refers to the 'always limited present, which measures the action of bodies as causes and the state of their mixtures in depth', whereas Aion refers to an ideal instant or 'empty present' that is infinitely divisible (LS: 61–5). This distinction can feel counter-intuitive. If the former is in one sense privileged as the 'physical' reality of a moving present ('only the present exists in time . . . [and] the present is in some manner corporeal' (LS: 162)), the latter allows for experience of change and meaning, precisely because Aion is limited to an ideal instant. That is, from the perspective of Chronos, what is, is in a forever moving present. But Aion reverses the order of priority such that it is the future and past that become the domain of meaning. That is, it is only future or past which matter, with the present being emptied into an 'instant without thickness and without extension': 'instead of a present which absorbs the past and future, a future and past divide the present at every instant and subdivide it ad infinitum into past and future, in both directions at once' (LS: 164).

This distinction between Chronos and Aion is crucial for cultivating an ecological attunement. The former incorporates the living present in which the becomings of actual occasions (Whitehead) or the passive syntheses of 'little selves' (Deleuze) simply are reality. But this living present is insufficient for understanding the experience(s) of purposive meaning. The eternal present of Chronos is not accessible to the limited vantage of the subject in motion, even if it can serve as an ideal reminder of cultivating equanimity. Moreover, if Chronos names the reality of a living present constituted by becomings, it also, in a certain sense, denies the viability of becomings in its very institution of an eternal present. Deleuze observes, for example, that

> whereas Chronos expressed the action of bodies and the creation of corporeal qualities, Aion is the locus of incorporeal events and of attributes which are distinct from qualities. Whereas Chronos was inseparable from the bodies which filled it out entirely as causes and matter, Aion is populated by effects which haunt it without ever filling it up. (LS: 165)

Aion thus becomes the crucial condition for the possibility of meaning even as this meaning is no longer 'given', but rather always an effect of one's orientation within events.

This point is both challenging and indispensable for an existential ecological attunement. We might say that selection/orientation are actions by which one fills out the emptiness of Aion in the hopes that one's attuning finds the most pregnant intensities in Chronos. Selection names the activity of meeting the event with an attitude that best allows us to 'become worthy of what happens to us, and thus to will and release the event' (LS: 149). But does this mean simply passively accepting what happens to us? On the contrary, when Deleuze says that 'willing the event is, primarily, to release its eternal truth' he invites us to understand the activity of selection as a positive creation of meaning. On the one hand, there is an acceptance insofar as events necessarily exceed our control, on the other hand, Aion is the time of meeting these events in a manner that best accommodates the greatest intensity of experience without becoming mired in resentments. This is not a bland or hopeless resignation or a facile positivity. It's not 'all good'. The selection involves understanding one's own role as a conductor between past and future, rather than solely as an outcome. This involves choice and creation, implicit in the activity of attention. Deleuze writes, 'we are faced with a volitional intuition and a transmutation' (LS: 149). The volition involves learning to 'will . . . not exactly what occurs, but something *in* that which occurs' (LS: 149).

Selection/orientation are two sides of ecological attuning as existential ideal. Selecting is the activity of becoming participant in the creation of meaning even as events exceed conscious control. What happens is one thing, it is question of selecting from the event its most intensely vital quality even in the midst of misfortune or suffering: 'my misfortune is present in all events, but also a splendor and brightness which dry up misfortune and which bring about that the event, once willed, is actualized on its most contracted point, on the cutting edge of an operation' (LS: 149). It is a matter of finding the cutting edge as an intensification rather than deadening of attention. This is a transmutation of equivalence. The forces of equivalence by contrast seek to dull the cutting edge through the churning operations of the Sameness machine. Without such selection, the path is already groomed by predictability. To select is thus to will that one not simply accepts the event insofar as its meaning is determined by a general or ready-made consensus and categorisation, but rather to enter into its 'communicating singularities effectively liberated from the limits of individuals and persons' (LS: 150). In this way, selection is a form of actualising the event's potential beyond prior identities or former categories; it is, as Deleuze will write later, 'a question of freeing life wherever it is imprisoned, or of tempting it into uncertain combat' (WIP: 171).

If selection is a manner of becoming worthy of what happens, orientation marks the manner or tone in which one incorporates this selection in anticipation. Selection/orientation are thus mutually implicated in all activities of attention. To attend is, whether consciously or not, to select and limit excess in order to achieve a coherent perspective. But, in the context of forces of equivalence and an increasingly loud media ever ready to supply pre-existing narratives for every event, then selection/orientation become both intensely important and increasingly imperilled. How do we select out of the noise to find that which is important or relevant? Even more provocatively, how do we create importance rather than simply follow the lines of what is created for us? Paradoxically, this creation of importance, this singularising of perspective, involves not a repetition of already consolidated narratives, but rather an openness to the possibility of these narratives being interrupted. Indeed, when Deleuze talks about learning to 'will' the event, he is inviting us to take up the reality of one's inextricability from events that exceed single ascriptions of their meanings even as they are the condition for the reality of subjectivity and its potential for authentic differentiation. Deleuze thus refers to a 'neutral splendor' that is 'impersonal and preindividual, beyond the general and particular, the collective and the private' (LS: 148). This 'neutral splendor' is crucial for an ecological attunement. Its neutrality is not a gesture towards a positivistic void, but rather an ecological expression of the interdependence between singularity and an ever-differentiating whole that does not stand still. This splendour exceeds any single fixed perspective on it, and yet it can be qualitatively impacted by any form of hegemony that levels difference. This is important especially insofar as so much of what is presented in the media as important may actually encourage a lack of attention or blindness to the qualitative specificity and singularity that is our unfolding ecology. Selection is thus also always a manner of anticipating what is to come. Can we learn to select with an orientation towards the creation of a future that does not repeat the abuses of the past while allowing for differences to proliferate?

Formally, Deleuze is consistently wary of how such selection is assimilated into a logic of negation. Ecological attunement requires an existential selection that does not insist on negating alternatives, but rather choosing which differences to affirm. An affirmation of differences involves developing or pushing their implications into a more intense state of actuality. In this sense, the selection is not simply a choice about how to best represent what really happened (though it is that as well), but rather is partially constitutive of actual occasions that emerge going forward. In moments of great intensity, this process appears automatic. But if differentiation and individuation are not only more complex than negation, but also prior to it, then even such intense moments of encounter are always full of possibilities that do not reduce to discrete oppositions. This is why Deleuze thinks of difference as affirmation, not negation: 'difference is the object of affirmation or affirmation itself . . . affirmation is itself difference' (DR: 52). Difference as affirmation requires the selection of which difference to affirm. To fail to do so is to be selected for and follow the given path of representation.

It is worth noting that Whitehead frames the question of selection in terms of evil, accepting evil as a necessary consequence of a fundamental temporal double bind:

> the ultimate evil in the temporal world is deeper than any specific evil. It lies in the fact that the past fades, that time is a 'perpetual perishing.' *Objectification involves elimination.* The present fact has not the past fact with it in any full immediacy . . . In the temporal world, it is the empirical fact that process entails loss: the past is present under an abstraction . . . The nature of evil is that the characters of things are mutually obstructive. Thus the depths of life require a process of *selection.* (PR: 340, emphases added)

This selection is a double-edged sword. It is 'at once the measure of evil, and the process of its evasion' (PR: 340). Whitehead orients a principle of selection that would develop sheer incompatibility into a patterned contrast and in so doing achieve a novel intensity. In this sense, the reaction to 'evil' is not a negation, but, similar to Deleuze, an affirmation of difference.

This insistence on affirmation rather than negation informs the second modality of an ecological attunement: its futural orientation. Importantly, such orientation is not teleological in a standard sense and cannot fixate on a prescribed outcome insofar as this will hinder the potential for the unexpected and prefigure or diminish attention. But how these possibilities are encountered and how the occasion is achieved depend to some extent on a basic affective attunement that orients towards the future. The tension is between a too-static generality and a too-restricted particularity. If an ecological attunement restricts general teleological prescriptions, it also is wary of restricting the orientation towards the future to conclusions or predictions attached to single perspectives. It therefore strives for a balanced openness and a willingness to be surprised. I have suggested that the conduction of past to the future through an ecological attunement is primarily mediated through affect and feeling. If our orientation of attention contributes to the unfolding of ecologies from which we are inextricable even as they exceed any singular perspective, then the best access we have to qualitatively influencing this unfolding is through the way we conduct affects and feelings. What are we attuned to notice? What feeling-tones do we collaboratively construct, create and pass on in our noticings?

This suggestion might appear to lead to a kind of facile optimism or neoliberal project in positive psychology in encouraging us to cultivate only 'positive' feelings. Worse, it might function to dampen or otherwise complicate the affective charges (anger, most predominantly, but also fear) correlative with critique of power or speaking out against injustice. Rosi Braidotti describes this as the 'relationship between creation and critique' constitutive of a central difficulty facing Deleuze-inspired political theory: 'how to balance the creative potential of critical thought with the dose of negative criticism and oppositional consciousness that such a stance necessarily entails?' (2011: 267). Everything depends on complicating superficial characterisations of the 'positive' or 'negative'.[21]

It is true that emphasis on creative becoming and affirmation of difference is often enlisted in rhetorics that emphasise the 'positive'. Braidotti herself for example describes the time of Aion as the 'time of becoming' in which 'spontaneous', 'creative', and 'joyful acts of transformation' invent new possibilities for the future (2011: 269). Whitehead also consistently emphasises the virtue of an attitudinal wonder rather than scepticism, suspicion or resentment. ('Philosophy begins in wonder. And, at the end, when philosophic thought has done its best, the wonder remains' (MT: 168).) Given the reality of risks, dangers and discords, we should take care to understand 'wonder' as normatively neutral and affectively heterogeneous. There is no guarantee that any occasion will only reveal 'positive' or joyful affects, nor that every encounter is respectful or reciprocal. This is a particularly important point in the context of the present, where harbingers of dire futures are increasingly the fabric of the everyday.

However, it is also because of this precipitous present that care must be taken in attuning to ways to avoid self-fulfilling prophecies. Undoubtedly this remains a delicate, ever delicate, issue. Indeed, in such a context, Deleuze's notion of the 'counter-actualization' by which one becomes the 'actor of one's own events' is crucial (LS: 150). Here, he again partakes of the Stoic ontological distinction between a realm of a cosmic physical whole and the 'incorporeals'. This is, in Whitehead's language, a distinction between God's consequent nature as the inclusive and incomplete present and the process of individuation of the singular occasion. The actor is the one who acts. In this sense, we are all, potentially, actors, but the question has to do with a capacity to be the agent of our actions and act in accordance with constitutive processes of individuation that are ontologically excessive to our ego identity. This is not about satisfying egoistic desires. It is, rather, attuning to the 'impersonal and preindividual':

'the actor strains his entire personality in a moment which is always further divisible to open himself up to the impersonal and preindividual' (LS: 150). In this way, the actor 'actualizes the event, but in a way which is entirely different from the actualization of the event in the depth of things . . . [because] the actor redoubles this cosmic, or physical actualization, in his own way' (LS: 150). This counter-actualisation is a manner of creating a meaning in relation to events. These events can be extreme, destabilising, even destructive. How do you counter-actualise them?

With this question in mind, we must insist that positive is not a characterisation of a facile optimism or celebration of 'happiness'. Far from it, especially insofar as images of happiness are frequently conflated with predictable security. And yet, there is still some conceptual valence to the distinction between the affirmative and the negative. Braidotti for example links this distinction in terms of its relational effects. The reinforcing feedback loop is important. Negative affects both tend to be generated by 'a blow, a shock, an act of violence, betrayal, trauma, or just intense boredom' and they tend to produce, inflict or perpetuate themselves through rigidifying the subject's capacity for relation: 'Negative passions do not merely destroy the self but also harm the self's capacity to relate to others' (Braidotti 2011: 288). By contrast, affects that increase the capacity to connect, relate, affect and be affected are deemed positive. The arbiter is thus relational capacity or affordance. This does not mean that all difficult feelings are negative.

It is always a question of when, how much, how, what manner. Indeed, Braidotti articulates ethical implications as 'the transformation of negative into positive passions, i.e. moving beyond the pain' (2011: 290). It is not a matter of ignoring or blocking out pain but of working through it. Such a working through is not necessarily a 'matter of retaliation or compensation but rather rests on active transformation of the negative' (Braidotti 2011: 293). This transformation is neither exclusively juridical nor epistemic, but ontological. The criteria is the effects on one's capacity for relation. The work of cultivating this capacity involves transforming negatively attuned reactionary modalities:

> We need to unlink pain from the epistemological obsession that results in the quest for meaning and move beyond, to the next stage . . . let us call it *amor fati*; we have to be worthy of what happens to us and rework it within an ethics of relation. Of course, repugnant and unbearable events do happen. Ethics consists, however, in reworking these events in the direction of positive relations. (Braidotti 2011: 293)

Rather than a cheap positivity based on denial, this is a deepened ontological capacity to find a way of selecting from the event that creates greater capacity for relation in going forward. It doesn't need to happen all at once. Grief, pain and sorrow are necessary components of the processing of trauma. But they are oriented towards a renewed capacity for relation, not as ends-in-themselves. In this sense, rather than a retreat into an idealistic false optimism, what Braidotti is expressing is a desire for a more resilient and sustainable critical stance. Indeed, Braidotti believes that emphasis on mourning and vulnerability cannot undo the pathological patterns of advanced capitalism. Equivalence runs on negative affects like fear and loathing. While there are, undoubtedly, many good reasons for the proliferation of such affects, it is a question of counter-actualising untapped vitality so as to create alternatives not pre-captured by equivalence. As Braidotti puts it, 'the politics of melancholia has become so dominant in our culture that it ends up functioning like a self-fulfilling prophecy', such that 'melancholic states and the rhetoric of lament [are] integral to the logic of advanced capitalism' (Braidotti 2011: 318–20).

It is also important to interrupt the assumption that the 'positive' relation is one that involves a certification of one's preconceived ideas, which, as we know, tends to arrive with a certain positive charge. But the issue is again ontological. Relation can be superficial if it is just a matter of 'connecting' through consolidation or agreement. By contrast, taking the full ontological notion of relation as part of the creative differentiation of reality means encountering difference and the unexpected. This is the difference between what Deleuze refers to as the 'lived states' of 'perceptions, affections, and opinions' and the ontological reality of percepts and affects and concepts unhooked from their consolidation in a representative consciousness (WIP: 170–5). If 'opinions are functions of lived experience' (WIP: 174) in its complacent ordinariness, it is a question of inducing the capacity to become other through the encounter with affect or percept prior to consolidation in opinion. In this way, living experience can be opened to potential transformation. By way of contrast, consider how social media networks often reinforce clichéd reactions. All sides of the political spectrum reinforce already constituted views through the sharing, tweeting and retweeting of images or articles that activate a feeling response that is already there: anger, scorn, fear, and so on. From an ontological perspective, the positive connection of this behaviour is superficial because it does not truly open to encounter at relational levels prior to one's already consolidated opinions.

For Deleuze, this is a manner in which art contributes to ecological attuning. 'The artist is always adding new varieties to the world' (WIP: 175) because they are working with affects rather than affections. In this sense, the artist is a creator of actual occasions that differ from the normal in unexpected ways.[22] But this differing, in a processual view, must remain to some extent open. Indeed, this is the difference between art and propaganda. Propaganda is directed towards reinforcing a message that is already there, whereas the intentionality of art stops at its affective and perceptive quality (tertiary). Nothing more and nothing less. While representative, thematic and conceptual content play a role, the point is not transmission of a view, but the engendering of an encounter and thus an actual occasion that exceeds the given.

These remarks need not be restricted to the humanly aesthetic. Each actual occasion is an achievement of reality that exceeds the perspectives of an habitual subject. The habitual human subject's access to this activity is through the modality of tertiary qualities, but such tertiary qualities are not ephemeral predicates of objects and subjects – they are reality! When fear happens, the actual occasion is fearful; fear has been actualised in the universe. When love happens, the actual occasion is loving; love has been manifested in the universe. Moreover, these affects/feelings become crucial to the further integration of the occasions to come. Debaise writes of a fearful animal:

> each action is inhabited by a modality of fear. It is the particular *manner* in which the past is integrated . . . Hence, that which is transferred from one act to another is not only the content of the act but the conditions by which a certain affective tonality ingress into a particular situation. While it always varies, intensifies, or, on the contrary, dissipates, fear is transmitted from act to act, forming the history of this particular route, which is the concern that has appeared in the life of this animal. (2017b: 70–1)

While it would be too simple to adopt any single-register normative assessment of feelings, the question for ecological attuning is again one of selection/orientation. How to select out of the feelings that which orients towards creation and relation? This does not mean pretending that there aren't real, and well-founded, fears. But it does mean cultivating attention to the manner in which feelings inflect the arrival of occasions and becoming reflective about which inflections increase our capacity to respond and which diminish or hinder this capacity.

In this sense, to speak of the positive or negative is already to risk reduction by encouraging stasis. It is always a question of how, the manner or style. A 'sad' affect is not necessarily negative if it induces an occasion of relational understanding. Nor is a 'joyful' affect necessarily positive in the broader sense: think of all the pain induced in the world by insistence on superficial happiness. Positive/negative is a simplification and abstraction from the complexity of qualities that resonate together in the achievement of an occasion. This requires careful attending to a distinction between the named content and the qualitative relation. We cannot characterise tertiary qualities as positive or negative before the fact. Positivity or negativity is a manner of what happens going forward, not the naming of a quality as such. It is contextual, relational and, moreover, open to further refinement and understanding as loops of experience unfold. In navigating the complicated webs and flows of material, discursive, ideological and political forces inflecting each occasion, ecological attuning means being open to unexpected convergences or the potential for new patterns of response.

The Janus-faced nature of time is such that no moment can be fixed or possessed. In this sense there is a perpetual processing of loss, as every actual occasion is an occasion of 'perishing'. There is no possession, it is always slipping through one's fingers. But, just as there is no secure possession of a fully determinate present, since as soon as the occasion is achieved it slips away into the past, so this past is also never fully eradicated. It lives on, informing the present, and this means both that the present is bound to the past, but also that the past is itself liable to new interpretations and new becomings as present achievements reshape its heritage. To fixate therefore on the loss of 'what was' is to miss the arrival of what might be. Correspondingly, to dwell only in the resentments of past tragedies is to risk repeating them in the present. This is not a celebration of ignorance or privileged denial. It is a question of degrees and dosage and an attention to unactualised potentials. Such attention is a response to forces of capture (whether political, ideological, pharmaceutical or economic) that would insist on the future proceeding along the same lines as the present.

Notes

1. This is the reason for the 'speculative ban' (see Hansen 2015a: 88–103 for a summary). Following such a ban, actual occasions are not phenomenologically experienced but rather are speculative constructions that function as heuristics for explaining experience. However, as I argue in Chapter 4, this does not mean there is no connection between lived experience and the speculative actual occasion, or in Deleuze's terms, intensive individuation processes.
2. Deleuze is interested in the ontological conditions of fatigue, which he locates in the 'natural contractile range' of an organism's 'contemplative souls' (DR: 77). This 'natural contractile range' correlates with the qualitative intensity of an actual occasion's positive and negative prehensions. In later texts, Deleuze associates 'exhaustion' with a choice to exhaust possibilities through dominance rather than open to new creation. This is the 'base' form of becoming that 'is nothing more than the will-to-dominate in the exhausted becoming of life' (TI: 141).
3. This formulation is from Fred Moten: '. . . you talk about being able to be in two places at the same time, but also to be able to be in two times at the same place . . . [this] is very much bound up with the Jamesian notion of the future in the present' (Harney and Moten 2013: 131).
4. The German term is 'die stimmung'. Though Macquarrie and Robinson render this as 'mood', the literal German connotation involves tuning a musical instrument. In

the English translation of *The Fundamental Concepts of Metaphysics*, it is translated as 'attunement'. See also Kuperus 2007 for a helpful secondary discussion.

5. I refer to Heidegger's three theses in the 1929/30 lecture course on the *Fundamental Concepts of Metaphysics*: '(i) the stone (material object) is *worldless*; (ii) the animal is *poor in world*; (iii) man is *world-forming*' (1995: 177). Though some have wondered if Dasein could apply to certain nonhuman animals, Heidegger's intent appears to be to use Dasein to get at something essential to the human being – '*We name the being of man being-there, Dasein*' (1995: 63). For discussions on expanding the scope of Dasein to nonhuman animals, see Thomson 2004: 401–3; Thomson draws on Haugeland 1982 in distinguishing in principle Dasein and the human being.

6. Because affect attunement 'emerges below the level of conscious awareness' (Willett 2014: 83), it might help describe the way affects move through groups. Willet suggests extending the notion of attunement from micro-organisms up to what is sometimes called a 'superorganism'. Drawing on recent research (Raison et al. 2010) which details the affective role that micro-organisms in the 'gut brain' play, Willet argues that attunement helps understand resonances that do not require a conscious subject: 'not just microorganisms, but massive superorganisms, as described by various social network theories, can likewise regulate the affect and physical function of nodes – aka people – through a process generally mysterious and yet statistically measurable. It is as though we humans not only have a multiply inhabited gut brain, but belong to a larger one' (2014: 84).

7. At least, if by ethical theory we presume that the obligation of such theory is to provide determinate decision procedures for ethical action, something I question in Chapter 4.

8. This possibility has a long philosophical history, as Stern notes in introducing it. Most notably, it dates back to Aristotle's affirmation of a 'common sensibility' (*koinê aisthêsis*) in explaining how the sense perceptions unite. (*De Anima* III.1 425a29; 1987: 187–8).

9. Sylvan Tomkins, an important source for Stern, distinguishes affects from drives. In contrast to affects, drives are object dependent (Sedgwick et al. 1995: 54–5). Tompkins's example is air: the *drive* for breathing is *only* satisfied by air. By contrast, the affect-object relationship is characterised by a high degree of variability and the causal relationship between affects and objects is neither unilateral nor rigidly linear. When an attuning transfer occurs, it does not have to express in the same way. Tompkins's affect theory thus better fits the phenomenon of subjective variance: 'Everyman has been puzzled for centuries at the irrationality of affect investment, that this one who has every reason in the world to be happy is miserable, whereas that one, whose lot is unrelieved misery, seems nonetheless to be full of zest for life' (Sedgwick et al. 1995: 54).

10. Because the logic of affect resists the logic of identity, it has been attractive to Deleuze-influenced theorists like Protevi 2009, Manning 2015 and Massumi 2002, all of whom draw on Stern and affect theory.

11. This does not mean that empathy is not an important pedagogical component to the development of an ecological attunement but only that empathy is in many ways a re-narration in more conventional categories of a metaphysical relation that is its condition of possibility.

12. There is some debate about the list of categorical affects. Sylvan Tomkins, an important source in affect theory, identified eight basic affective complexes: interest-excitement; enjoyment-joy; surprise-startle; distress-anguish; fear-terror; shame-humiliation; contempt-disgust, anger-rage (Sedgwick et al. 1995: 74).

13. This follows the discussion of Uexküll in the Introduction.

14. Alice Jardine reads this as one way in which Deleuze's theory remains inscribed in a masculinist logic and fails to develop its more radical potentialities. See Jardine 1984.

15. The appendixes of *Logic of Sense* gather five previously published articles in slightly modified form. The relevant appendix ('A Theory of the Other'), originally published

in *Critique* in 1967, is structured around a literary analysis of Michel Tournier's novel *Friday* and the concept of perversion. As such, its aim is different than the more metaphysical *Difference and Repetition*. However, the two accounts are consistent and mutually reinforcing.

16. I am convinced by Jardine's criticism of the terminology of 'becoming-woman'. Even if, as she acknowledges, the concept is not meant to refer to empirical women, it is likely to be misunderstood and easily read through problematic gender norms. However, for the a priori Other structure presented in the *Logic of Sense*, Jardine's criticisms primarily focus on Deleuze's reading of Tournier and the concept of perversion. I am more interested in the metaphysics of the structure, which need not be read through a psychoanalytic lens or in the context of gender or sexuality per se.

17. Zourabichvili observes that the fundamental logical movement of Deleuze's philosophy is 'implication' not explication. To explicate is to separate or translate an expression of intensity into description in terms of qualities and extension. To implicate, by contrast, is to fold two series or centres of resonance together, while maintaining their difference (2012: 105).

18. Deleuze offers two different descriptions of the metaphysics of time: the tripartite analysis in *Difference and Repetition* and the Chronos/Aion distinction in *The Logic of Sense* (this latter distinction returns in the Cinema texts). While there is brief reference to Aion in DR, he does not provide any analysis of how these two descriptions relate to one another. I read them as differing primarily in terms of motivating problem: the three syntheses in DR respond to how a transcendental empiricism accounts for time without recourse to the transcendental ego. The Chronos/Aion discussion in LS is concerned with existential and ethical implications. I focus on the latter above.

19. Deleuze is interested in the Stoic ontology as an early alternative to Platonism (the Ideas are no longer dominant over bodies, but rather in some sense the 'effects' of bodies) and because of how these 'incorporeals', in particular the 'sayable', bring into view the problem of sense. How is sense generated? Deleuze writes, 'one always returns to this problem, this immaculate conception, being the passage from sterility to genesis' (LS: 97). The sterility is states of affairs as they are, no more, no less. In themselves, there is no sense. And yet sense is where the bulk of our conscious energy is expended, even as it moves us to act on bodies. For a helpful secondary discussion of Deleuze's interest in Stoic ontology, see Bowden 2011: 15–47.

20. This leads to the familiar Stoic ethic. Given that, as Alexander writes: 'all things are bound together, and neither does anything happen in the world such that something else does not unconditionally follow from it or become causally attached to it' (cited in Bowden 2011: 20), one must develop a capacity for acceptance of this order.

21. It is worth distinguishing between Deleuze and Deleuze-inspired readers. In this context, Andrew Culp's polemical intervention *Dark Deleuze* charges Deleuze scholarship with establishing a 'canon of joy' that falsely portrays Deleuze's work as 'a naively affirmative thinking of connectivity' (2016: 2, 65).

22. This language is misleading since it makes affects sound like completed things. In English it works better to discuss this in terms of *affect-ings* as opposed to affections. The affect-ing is asubjective and inherently relational, the affection 'belongs to' a subject as a lived state.

Conclusion:
Fabulation and Epoch(s)
to Come

Attention occupies a paradoxical existential function, both open to influence by wilful selection but also by degrees creative *of* will and control. However, if attention is potentially constitutive rather than passively receptive, this result alone does not suffice to actualise its creative potential. In what follows, I articulate important implications for an ecological attunement endeavouring to maximise such actualisation and then situate these implications in the context of ecological crisis and the future.

Implications for ecological attunement

There is a risk that 'ecological attunement', as a concept, appears only as what Jonathan Lear has called an 'enigmatic signifier', a vague and underdetermined gesture towards some presumably beneficial or more harmonious manner of being (Lear 2000).[1] This risk is not however a matter of a simple error. That is, our response cannot simply be a more concerted effort at explicit *definition* of the concept to remove all uncertainty. Such an approach could only in this case reduce the concept's power for effecting living experience. This dilemma follows the perspectivism I have argued for throughout – there cannot be one manner of ecological attunement. To say there was would be unecological.

This does not mean that nothing can be said, but rather that my approach is marginal or lateral; rather than strike for the centre of the concept, I work around its edges, reviewing impacts or induced effects. As I have argued (Chapters 4 and 5), one effect is an imperative to approach what is presumed ordinary or normal from a different angle. Because no single perspective is co-extensive with the whole, different attendings are not best measured or evaluated primarily by their distance from a presumed norm. Rather, different attendings are potentially important sources of information about reality in its unfolding. Differently attuned relations have value in disclosing aspects of reality hidden to others.

In existential or lived dimensions, this imperative is not primarily epistemic. The obligation is not that all formulated perspectives (opinions) have equal validity. (Indeed, by the time we get to opinion, we are rarely attending.) The call is rather to direct attention towards edges and margins and places of transition. A form/content distinction is also important since margins are perspective-relative *in content*. The imperative is thus to attend to outliers or differences from the present norm of the attuner. This will manifest in differing ways relative to the 'historic route' of occasions constituting the perspective in question. There is in principle no limit to this imperative, though in practice it requires much discernment. If not every outlier will prove consequential, becoming more attentive to them remains crucial as a corrective to dominating tendencies towards their erasure.

In the context of customary connotations of 'ecology', a significant norm to be undone is the way the fundamental human/nonhuman dichotomy structures attention.[2] The urban professional walks to their car while all around them the crows squawk – for what and why? A dog's intense olfactory powers and their sensations of the Earth beneath their paws place them in contact with realities unperceivable by the human. Can an ecological attunement attend to what Deleuze calls '*interspecies junction points*' (WIP: 185)? Such juncture points are situated in individual encounters. While it is tempting (conceptually) to jump to the level of the 'whole' and affirm or celebrate an essential interdependence, this move, while certainly 'correct' in one sense, can counter-intuitively diminish attention to *differencing as it happens*. Instead junctions or encounters where different forms of feeling meet (the shoreline, the liminal, between 'species') promise qualitative insights.

This attending can be a decentring of the human, distancing ecological attunement from the notion of the human as uniquely 'world-forming'. To the extent that the human's constituted world is 'only' human, then it is ontologically impoverished. That is, it is a narrower and more reductive selection from the real. Because this real exceeds singular perspectives even as it is also constituted by them, one's 'existential territory' is made richer insofar as it encounters a wider range of perspective and forms of life. The risk though is that this sounds optional, as if one chooses to be more inclusive to reap the benefits of a richer existential territory. Emphasising the metaphysical reasons for attunement shows the extent to which it is not a supererogatory ethical ideal, but rather ontological in source and implications.

Decentring the human follows an understanding of 'subjectivity' not based in a single sovereign essence, but rather constituted by processes interacting at myriad levels: from the biophysical to the affective to the ideological/conceptual to the social to the imaginary to the historical, and so on. One consequence is that one cannot presume that intentional understandings of present actions are exhaustive of their effects in reality. Conceptually, this is expressed in multiples registers of meaning and levels of causalities. This multiplicity of causality is essential to ecological thinking (Chapter 4) as well as how individuation processes emerge (Chapter 2). As evolutionary biologist Ernst Mayr asked:

> Why did the warbler on my summer place in New Hampshire start his southward migration on the night of the 25th of August? . . . I can list four equally legitimate causes for this migration. 1) An ecological cause . . . 2) A genetic cause . . . 3) An intrinsic physiological cause . . . 4) An extrinsic physiological cause. (1961: 1502–3)

How do we encounter such complexity of causalities without reducing them or mandating a selection of one as the real 'true cause'?

Whitehead's challenge to simple location enables a better thinking of this complexity insofar as there is no simple location or single temporal metric metaphysically. The fallacy of simple location is a mistaking of an abstraction for the real. The abstraction is the idea of a *simple* location, whether spatial or temporal. Following this conceptual critique, the existential implication is that the place/time where 'we' are is intensively complex and relationally promiscuous. A decentring of the human is an invitation to recognise this temporal and spatial complexity. Consider a classroom lecture as a complex intersection between series of occasions constituting students, lecturer, and so on. The lecture is constituted through ideological repetitions and transferences, previous lectures (both heard or given), texts, and so on that are re-presented, altered, transformed. These ideas themselves flow out into the room as expressed in spoken words (waves of sound) and images and there they encounter in different ways the psychic systems of listeners where they are filtered through histories of resonances and affect: previous concepts, hopes, fears and dreams. In addition, the lecture room includes nonhuman creatures (ants, spiders, bugs of other sorts, as well as micro-biotic organisms) whose presence may seem negligible, but who are nevertheless part of that controlled ecology. This ecology is regulated through heating systems, lighting systems, concrete, wood, plastics all conspiring together in complex series of occasions that link this contained environment to wider networks of energy production and resource extraction.

The complexity of causalities belies the completeness of single perspectives. This is a descriptive claim. Temporal heterogeneity is lived through differing metabolic rates and perceptual affordances. Because of differences in rhythm, what looks fast to a human may appear slower to, for example, a hummingbird. Sound is a good example. The soundscape ecologist Bernie Krause shows how slowing down frequencies that are inaudible to the human ear reveals expressive properties of entities commonly thought of as mute, in this case cottonwood trees (Krause 2009).[3] Temporal and spatial perceptions are relative to the scales and rates at which one is 'pitched' and cannot be taken as unequivocally exhaustive of reality as such.

The existential decentring inherent in ecological attunement is more than simply recognising limitations or finitude or a multiplicity of perspectives. Rather, the relational quality of a processual reality has a positive implication insofar as it encourages cultivating a skill to modify the centre-periphery relation. One way of doing this is through background-foreground shifts. Discussing evolution, Gary Snyder offers a wonderful example of the merits of such gestalt play:

> the common conception of evolution is that of competing species running a sort of race through time on planet earth, all on the same running field, some dropping out, some flagging, some victoriously in front. If the background and foreground are reversed, and we look at it from the side of the 'conditions' and their creative possibilities, we can see these multitudes of interactions through hundreds of other eyes. We could say a food brings a form into existence. Huckleberries and salmon call for bears, the clouds of plankton of the North Pacific call for salmon, and salmon call for seals and thus orcas. (Snyder 1990: 117)

Snyder's suggestion, though drawn from different conceptual traditions, resonates with a processual manner of thinking individuation in exhibiting how such individuation does not proceed according to a linear efficient causality. It is this back and forth between perspectives without centring one as dominant that must be encouraged in ecological attunement. This is undoubtedly a lived challenge as it simply is the case that 'our' perspective is more intensely implicated for all but the saintly. But the goal is cultivating this perspective as more expansive and attuned to its own enmeshment in proliferating differences. This enmeshment is real, whether the subject is conscious of it or not. And it extends beyond the exclusively human.

Indeed, if foreground/background gestalt play indicates the extent to which perceptual complexity can inspire different modes of attention, the full implication for an ecological attunement goes further. When the phenomenological structure of attention fixes one polarity as given (the attend-er), it encourages a habitual blindness. By contrast, a processual conception of subjectivity intensifies an inner bipolarity – the subject is never only subject, but always also object. This is to say that the subject is in relation to other subjects, but also that this subject/object dialectic *is the pulse of reality.* Deleuze puts this point nicely in writing about Whitehead:

> Everything prehends its antecedents and its concomitants and, by degrees, prehends a world. The eye is a prehension of light. Living beings prehend water, soil, carbon, and salts. At a given moment the pyramid prehends Napoleon's soldiers (forty centuries are contemplating us), and inversely. (1993: 78)[4]

The pyramid's prehension of Napoleon's soldier, the prehension of time immemorial in the encounter with the redwood tree. When we touch, taste, see, feel, we are being touched, seen and felt. This is ecological attuning.

If such prehensions (great pyramid prehending soldiers, tree prehending human) are often overlooked, that is because more immediately vexed prehensions are between person and person, between state and state, between institution and person, and so on (not to mention between animal-food-prey and predator-eater). These more immediately vexed prehensions, regardless of metaphysics, tend to occupy our whole attention. While we can conceptually understand temporal heterogeneity, relational complexity and the fallacy of simple location, we still nonetheless must act. This necessity presents itself to us in contexts that appear confined to a present situation. Indeed, insofar as action is possible, the situation must be determined as what it is, and its relations to other situations, times, places, and so on, must be demarcated and limited. We cannot then espouse ecological attunement in terms of a union with the 'whole', or, as has so often been the case, motivate our actions through an appeal to a judgement deemed correlative with the whole.

And yet, is there not still a sense in which we should endeavour to feel and act, as much as possible, with an ideal conception of a 'whole' that exceeds a more limited sense of self-interest? Better put, can we learn to understand that our perspective, if it is to be vital, is a singularisation of relational complexity that extends throughout the universe, while at the same time, recognise that we do not 'know' this complexity in a fully determinate propositional manner? This involves nothing less than a radical destabilisation of deeply embedded habits of judgement. In the ruptures opened by such destabilisation, what can serve in their stead? This is where 'feeling-tone' and affect become the closest access we have to the processes of the real that exceed categorical classifications. This does not mean that they function as unequivocal source of judgement. This is the difficulty. And yet, though the shift is subtle, it nevertheless is real in impacting how we think about encounters and actions. Indeed, to encounter and act is always in terms of feeling-tones engendered, received, transferred or cultivated. Transforming habits, behaviours and actions in a real and lasting way must work at the level of feeling as constitutive of habits.

This does not mean relinquishing all judgement. Judgement remains a form of motivating action. As Michel Serres reminds us, the need for judgement comes from the persistent threat of violence and death.[5] This threat does not disappear with an ecological attunement. But, following Serres, the question is how to use reason so that the necessity of judgement does not become merely an arbitrary caprice of power. Reason 'comes from fidelity to the real' (Serres 1995: 92), and in this sense what I have offered in this text is a reasoned argument about how such fidelity leads to ecological attunement.

For such an attunement, living access to the 'whole' is through 'feeling-tone' even as such feeling-tones are singularised and hence *understood* from a perspective that is not equivalent to the whole. The distinction between the feeling and the understanding of the feeling is where the discernment of experience lies, and ecological attuning is nothing if not a call to better attend to this gap. We must remain wary of an appeal to the 'whole' from the interests of one perspective and an image of union with the whole that is merely a projection from consolidated identity. Both of these misconceptions reinforce each other and impede attention to difference or junctures across which feeling-tone manifests. Indeed, rather than reinforcing a perspective, the whole or the 'universe' is that which has the power to *destabilise* an existential location, not that with which a meaningful union can be sought. Deleuze demonstrates this in discussing the relation between 'house' and 'universe'. The universe-cosmos is always in relation with the instituted territory of the organism, but this territory or 'house' is not simply a part or segment of a larger whole. Rather, just as a territory or house is composed of patterns of sensations and affects that represent what Deleuze calls its 'endosensations', it is also susceptible to the maximum force of disintegration – the universe: 'the territory does not merely isolate and join but opens onto cosmic forces that arise from within or come from outside and renders their effect on the inhabitant perceptible' (WIP: 185–6). Such a relation is tenuous, since the universe-cosmos, of necessity, exceeds the territory or house, even as it is in filtering, shifting and giving form to these universal-cosmic forces that the territory makes a spatio-temporal habitat for the subject.

An ecological attunement dwells at the uneasy intersection between existential territory and universe-cosmos as maximal gesture towards an open Infinite. If an existential subject is built out of patterns of feeling it is always potentially in relation with forces that can eradicate these patterns – the universe as cosmic 'deterritorialisation', not harmonious whole. It is precisely this uneasy intersection that gives rise to the compulsion of equivalence as a means to deny it. But it also at this uneasy intersection that attention to manners and forms of feeling of the Other become affirmative modes of resisting the tyranny of equivalence. At their best, the affirmative ontological potentials of such attention may be creative of occasions that escape equivalence. As Jean-Luc Nancy writes,

> it is a question of a particular consideration, of attention and tension, of respect, even of what we can go so far as to call adoration, directed at singularity as such . . . [this is] an esteem in the most intense sense of the word: a sense that turns it back on 'valuation' measures. Esteem . . . summons the singular and its singular way to come into presence – flower, face, or tone. (2015: 39)

Fabulation and imagination as affirmative resistance

Faced with a present thick with portentous omens, appeal for attention to the singular may feel negligible, even irrelevant. While Nancy is talking about flowers, we have rising sea levels and temperatures, accelerating extinction rates, unprecedented numbers of displaced people, the looming potential of resource wars and the rising rhetoric of right-wing nationalism across the globe. Blithe denial of such omens will ensure that ecological attunement remains only a fantasy of the fortunate or sequestered. Nevertheless, in the context of such portents the intersection of affect, attention and ontological responsibility intensifies: how to live affirmatively in a present teetering under a parade of dystopic and apocalyptic presentiments?

In adopting a primary focus on the psychic, the worry is that this forecloses attention to the inter-psychic or social. Though a processual conception shows that no psychic individuation is effectively separable from cascading series of relations far exceeding the individual, given the extent to which this result remains speculative, the worry retains a degree of purchase. Full engagement with the social and material dimensions of these portents was not the goal here, but there are some relevant implications for an ecological attunement to consider as closing provocation.

In such contexts of deep fear and uncertainty, one difficulty is cultivating awareness without succumbing to paralysis, wilful ignorance or the oppositional bifurcations that so frequently accompany popular thinking. The most basic bifurcation is a pervasive temptation: 'us versus them'. Undoubtedly such a bifurcation is a powerful motivator of human behaviour. It remains, moreover, fundamental to metaphysical accounts that rely on negation as the necessary condition of determination. For an ecological attunement following Deleuze and Whitehead, however, contradiction or oppositional negation does not represent a metaphysical necessity for determination, but rather a frequently trivial effect:

> It is not difference which presupposes opposition but opposition which presupposes difference . . . Our claim is not only that difference in itself is not 'already' contradiction, but that it cannot be reduced or traced back to contradiction, since the latter is *not more but less profound than difference.* (DR: 51, emphasis added)

How does this metaphysical understanding alter lived orientations, especially in a context where so much of reality unfolds through violence or opposition?

The 'us v. them' bifurcation is foundational for societies tending towards closure since their method of survival involves selecting as relevant from occasions only that which perpetuates a narrowly repeatable pattern. By contrast, an open society is less liable to insist on such a bifurcation because its strategy for perseverance is one of creative adaptation. In this way, such an open society is always becoming different. Undoubtedly this is a delicate point, especially in contexts of violence and reactionary sentiment. To valorise becoming different as always desirable would be too naïve; doesn't it matter how and in what direction such becoming proceeds? It is easy to say that from a metaphysical perspective all patterns of identity are transient or ephemeral, it is even easy, perhaps, to *know* this – but what does it mean to live it?

In particular, what does it mean to remain open when faced with encounters with others whose views you find odious or who have a real intention of harming you? Ecological attunement does not mean an obligation to tolerate, accept or otherwise remain passive in encounters with differences manifesting in this manner. If someone is chasing you down the street with a knife, you run – as Frank O'Hara once said when asked about his poetics.[6] But that is not the end of the story. While no philosophy can replace the need for existential discernment, different philosophies offer different tools or suggestions for such discernment. We can keep the open/closed distinction as conceptually fundamental for ecological attunement while insisting these distinctions are by degrees and there is no determinate inoculation against the uncertainty of becoming for a society oriented towards 'openness'. But such an orientation has implications for how one approaches conflicts and challenges. For one, understanding the role of affect as ontologically constitutive leads to a different manner of encountering different narratives or views. This does not mean an epistemic free-for-all; not all narratives are equally valid or worthy. Nonetheless, if one is truly interested in persuasion as opposed to force, then one must endeavour to understand the affective economy that is being enlisted in a particular narrative or view.[7] What are the fears or hopes which drive a particular realisation or actualisation? In conditions of difference, genuine rather than coercive transformation can only happen when the feelings constituting the other's reality shift. Reasoning has a role to play, but reason is most effective when it understands the constitutive role of feeling and works at this level.

What does it mean to work at the level of feeling, where affective variability is the rule not the exception? Since feeling does not automatically translate into universal normative judgements the discernment of ecological attunement remains provisional. Certain occasions of fear are more reliable than others, for example, for motivating action and certain responses to fear will be more effective than others in relation to the context and cause. The feeling of fear is nonetheless real, and it is there that one begins. Ecological attunement includes reflection that returns to 'select' (Chapter 6) from experience, that dwells with experience, that considers past experience in light of subsequent understanding. Attention in the present is thus shaped by this reflection on the past as well as orientation towards images of the future.

A criterion for such selection can be offered in terms of connection: how does an attention contribute to creating connection? How does an action contribute, or not, to opening more relational capacity? Such a question provides ballast to Deleuze's 'line of flight' that destabilises, problematises or dissolves any consolidated mode of identity. It is a question of attuning to the difference between a blockage that sediments and an opening that leads to richer qualitative connections. Not all 'lines of flight' are positive – they can lead as easily to death and isolation as to more intense capacities for relational experience – but sometimes, especially in contexts that seek to predetermine or control outcomes in the name of equivalence, a line of flight can be a means of creative invention through creating connections.

It remains a question of 'dosage' which cannot appeal to a static categorical judgement that always applies in the same way. And yet, it is also the case that selection must be made. Every actual occasion involves both positive prehension (which includes but also transforms) as well as negative prehension, which omits or reduces. As we have seen, for Whitehead, the processual nature of reality itself exhibits an irremediable 'evil' ('In the temporal world, it is empirical fact that process entails loss' (PR: 340), Chapter 6). The actual world is, by its very nature, impermanent. Nothing in actuality stands still, nothing in actuality remains forever, the world of actual occasions is a world of perpetual loss.

However, in doing away with enduring substances, Whitehead's philosophy paradoxically opens a different orientation towards permanence. What is permanent is not an object or subject that remains unchanged, but rather each moment of reality as it passes into an objective immortality leaving its trace on what is to come, even as it also expresses previous traces. The response to the suffering of finitude is not to somehow seek to become infinite *as one is*, but rather to understand that the condition of limitation is required for anything to become actual. It becomes a pragmatics of selection and a pragmatics of feeling. What feelings connect with what other feelings, what leads to what? Whitehead writes: 'A feeling bears on itself the scars of its birth; it recollects as a subjective emotion its struggle for existence; it retains the impress of what it might have been, but is not' (PR: 226–7). The question of beauty as telos (Whitehead), or intensity of experience (Deleuze), inheres in a capacity for transforming apparently intractable oppositions into more capacious or complex patterns of difference.

Indeed, in a closet-Nietzscheanism, Whitehead declares that both evil and good are positive in themselves – their difference lies in what they enable: 'Evil is positive and destructive; what is good is positive and creative' (RM: 96). Can we learn to notice what attendings and selections create rather than destroy while nonetheless admitting that this does not always map onto received moral categories? Indeed, Whitehead is explicit about the dangers of inattentive or complacent moralisms: 'Good people of narrow sympathies are apt to be unfeeling and unprogressive, enjoying their egotistical goodness . . . this type of moral correctitude is . . . so like evil that the distinction is trivial' (RM: 98).

Since all selection is at once inclusion and exclusion, another risk is existential paralysis generated by a desire to avoid any limitation or exclusion. (This is precisely what Deleuze calls the 'greatest danger' for a philosophy of difference (DR: xx), that of the 'beautiful soul' who avoids the necessity of choice.) In a context of IWC and equivalence, this paralysis, whether willingly or not, functions primarily in the service of the status quo, Melville's Bartleby notwithstanding.[8] The very nature of reality *requires* existential selection: 'the depths of life require a process of selection' that is a double-edged sword 'at once the measure of evil, and the process of its evasion' (PR: 340). The question remains: how to make selections without knowing for sure? How to choose openness and uncertainty rather than fixation on a perceived certainty?

This question is intensified by the metaphysics in question, precisely because choices, actions and attention matter. For both Whitehead and Deleuze, the transformation of an apparent opposition or negation into a greater complexity of patterned contrast is how novel intensities occur. This does not mean that they just happen. By contrast, our actions, choices and attendings become more implicated in the creation of reality even as this creation remains in principle uncertain from any determinate present vantage. Metaphysically, novel intensities do not necessarily have to be 'good' from a human perspective. Stengers observes: 'Whitehead's cavalier perspective . . . *is neither that of a righter of wrongs nor of a denouncer*' (Stengers 2011: 292, emphasis added). This is because 'creativity, Whitehead's ultimate, is neutral with regard to the distinction between cosmos and "disordered" multiplicity' (Stengers 2011: 308).

Because we cannot be certain how to proceed towards open creation with a guarantee that it will map onto present understanding of a desired outcome, the closure of a determinate and static identity becomes a particular temptation in times of duress. But it is also under such conditions that the need for creative response is most significant. In the context of this tension, Deleuze's repurposing of Bergson's concept of 'fabulation' is provocative. For Bergson, fabulation is primarily a negative operation within closed societies that seeks to maintain an illusion of stable identity by perpetuating an image of an ideal past. Fabulation for the closed society is always in terms of what was (Bergson 1935; Bogue 2007).[9] Deleuze repurposes 'fabulation' as a means for open societies to disrupt the stultification of orthodoxy and forces of closure. Indeed, insistence on static identity and its correlative 'us v. them' logic is always accompanied by a hegemonic claim of access to the real. The 'us' are those who know; the them are those befogged, befuddled or otherwise just wrong. There is a clear and determinate distinction between what is and what is merely imaginary. By contrast, Deleuze's conception of fabulation and what he provocatively terms 'the power of the false' operates at the 'frontier between the real and the fictional' (1989: 153) in recognising the power of fictions as actual forces in the world.[10] In this sense, his notion of 'fabulation' has some similarity with what Whitehead calls 'adventure'.

In both cases, the concept is not epistemic, despite caricatures of so-called 'postmodern' evacuation of truth. Instead, fabulation functions at the edge of the present real and the future imaginary. It is (one of) the ways by which alternatives emerge. Fabulation in an open society brings into view alternatives for encountering events that do not fit a predeterminate scheme. If in a closed society fabulation celebrates static illusions as source of identity, in an open society fabulation is a feature of improvisation. It is oriented towards the future and is a means that allows for a 'people to come'. But the great rub is that it cannot do so prescriptively. Fabulation is not a fully determinate conceptual representation of an alternative future that we then endeavour to create. Such a representation may be a useful inspiration, but if it becomes too static it risks becoming another form of closure, a representative model that blocks rather than fosters becoming.

The great paradox of fabulation/adventure is that it runs ahead of that which can be safely conceptualised and represented within the language of its time.[11] It acts as a force from a future that is not yet presently articulable, even as it disrupts the habits of that present. There is a necessary element of risk involved in such activities since they are frequently deemed fanciful from the standpoint of those for whom complacent habits have been reified into structures of what 'really is'. Such reification leads to blockage or breakdown of the ability to think beyond what is presumed to be given. Moreover, what is presumed 'given', obvious, real, indisputable, and so on, often appears differently from different perspectives. No perspective is immune to this potential attachment to a presumed givenness that blocks fabulation rather than encourages it. This is not however to make the whimsical claim that fabulation/adventure is without constraint. We do not just make up whatever future we want, whatever suits our fancy.

This is where the turn to affect and what Whitehead calls a 'lure for feeling' becomes so important. The lure for feeling functions as 'the final cause guiding the concresence of feelings' (PR: 185). It is a living expression of the hybrid between potential and actual, what might be as it is felt, not known. But, if the promise of fabulation or adventure lies in disrupting sedimented habits to open gaps for different attunements, how can such disruption proceed without overdetermining its outcomes? In the context of a processual openness, fabulation cannot be a determinately represented alternative. That is called propaganda, strategic manipulation whereby one invents narratives necessary to suit antecedent purposes. This presumes the stability of purposes unchanged despite changing conditions. We cannot definitively prescribe a future; the future will look different than expected from any single vantage. This is axiomatic for a process philosophy. This is why the logic of equivalence, in its manner of bleeding from the economic into the epistemic and social, is so dangerous, because it predetermines what counts as a value and blocks the coherence of novel forms of value emerging (or re-emerging). In doing so, the mere activity of imagining a future that does not proceed according to behaviour dictated only by profit-seeking is made to seem wildly implausible, unrealistic, naively idealistic. This ends up a self-fulfilling prophecy in the most dystopic sense.

This is a particularly charged tension in the context of a present dominated by ominous harbingers. How can we affirmatively foster counter-images that are not head-in-the-sand delusions? This question is isomorphic with the way that perceived images of the future inform becoming in the present. Faced with ongoing turbulence and the likelihood of intensifying destabilisation, the closed society doubles down on preserving an identity presumed fixed. The open society seeks to experiment with means of becoming worthy to the intensity of events of deep change.

A difficulty of this distinction is the temptation to conflate such choices with ideological content. All ideological content risks becoming a form of closure if we are not attentive to the forms in which its ideals or norms manifest. This doesn't mean of course that there is no place for critique or contexts in which direct action is necessary, but it is to say that universal mandates of such critique or action are unlikely to truly foster creative encounters that engender true transformation. The negative or affirmative tension need not always play out as correlative with substantive or fixed ideological positions. This remains a challenging issue for existential discernment. But its implication can be summarised as an effort to understand conflicts and dissonance as opportunities for unexpected creations, while likewise understanding that creativity is not a magic word. It is a descriptive term, not a normative one. Actual occasions continue to happen, and reality continues to be created in ways excessive to any singular perspective. Rather than orienting simply towards 'being right' and negating those who are 'wrong', a task for futural fabulation is seeking how to make unexpected connections and thus create alternative territories where new modes of flourishing become possible. Sometimes this may involve relinquishing one's attachment to a perceived identity, or, at least, understanding how that identity is open to reformulation under different conditions and in different relational contexts.

If we cannot *know* what reality will become, we can go there moment by moment through careful attending to the tertiary qualities that enrich, deepen or intensify our sense of its complexity. Or we can recoil in fear and understand the challenges of change through a lens of diminishing returns, thereby fixating in a manner that preserves habits constitutive of ecological crisis itself. While an ecological attunement relinquishes the dream of absolute mastery or control, it also simultaneously heightens responsibility to one's role as collaborator in the continued achievement of occasions of reality. The challenge is in understanding how to fit these two together by both letting go (of the blindness of certainty) and becoming perspicuous in attention to the present (without presuming to know what it means). As Deleuze writes:

> A concert is being performed tonight. It is the event. Vibrations of sound disperse, periodic movements go through space with their harmonics or submultiples. The sounds have inner qualities of height, intensity, and timbre. The sources of the sounds, instrumental or vocal, are not content only to send the sounds out: each one perceives its own and perceives the others while perceiving its own. (1993: 80)

We are already participating in this cosmic concert – it is a question of learning to hear and cultivate more attuned responsiveness.

Notes

1. Lear develops his notion of 'enigmatic signifier' in psychoanalytically inspired reading of Aristotle. Lear is interested in how appeal to 'the good' at the beginning of the *Nichomachean Ethics* functions as such a signifier in the interest of an 'inaugural instantiation': '[Aristotle] is attempting to inject the concept of the "the good" into our lives – and he thereby changes our lives by changing our life with concepts' (2000: 8). The enigmatic signifier is part of a process of creation. Rather than defining a concept that already exists, Aristotle is 'inducting us into a new way of life' (2000: 9).

2. Erin Manning's exploration of how a process metaphysics offers a different epistemic framework for thinking about neurodiversity, in particular autism, is also relevant to this point (Manning 2013).

3. For a short excerpt discussion of this taken from a longer film, see: https://www.youtube.com/watch?v=uWkMWDSVZuQ. For the complete film, see: http://library.fora.tv/2009/09/22/Dr_Bernie_Krause_The_Great_Animal_Orchestra

4. Deleuze's use of Whitehead's technical term 'prehend' is loose. Technically, actual occasions prehend but all of the examples given are 'societies'. That said, the distinction here is pedantic if used to dismiss Deleuze's basic insight, which is consistent with Whitehead.

5. Serres distinguishes reason and judgement. Where reason 'presides over knowledge and science', judgement 'presides over legal reason [and] the need for arbitrating'. This need 'comes from violence and death. Without the arbiter, we would be exposed to worse risks, we would kill one another' (1995: 92).

6. See 'Personism: A Manifesto' in O'Hara 1976.

7. Whitehead suggests in *Adventure of Ideas* that one mark of civilisation is the extent to which persuasion replaces force. To persuade someone who does not see things in the way that you think they should be seen requires you to understand how and why their vision is constituting the narrative it does. Whitehead's notion of persuasion does not assume that facts are self-certifying or that their orthodox interpretation is all that is necessary. For investigation of this notion of persuasion in the context of political theory, see Dombrowski 2017.

8. While I will not pursue this here, Bartleby is not paralysed, rather Bartleby's 'I would prefer not to' can be read as a positive resistance.

9. Bergson's account of 'fabulation' is developed in chapter 2 of *The Two Sources of Morality and Religion*, where he aligns it with how religion in a closed society reinforces the status quo through appeal to gods and mythical ages of past greatness. In the English translation, 'fabulation' is rendered as 'myth-making'. See Bergson 1935: 83–178.

10. For an excellent account of Deleuze's alternative deployment of 'fabulation', see Bogue 2007. For investigation of the risks of such rhetoric in the context of 'fake news', see Duvernoy 2020.

11. To help explain adventure, Whitehead uses the example of a scholar who too reverentially privileges the canonical past: 'the definition of culture as the knowledge of the best that has been said and done, is so dangerous by reason of its omission. It omits the great fact that in their day the great achievements of the past were the adventures of the past. Only the adventurous can understand the greatness of the past . . . To read [the classics] without any sense of new ways of understanding the world and of savouring its emotions is to miss the vividness which constitutes their whole value' (AI: 279).

References

Allan, George. 2008. 'A Functionalist Reinterpretation of Whitehead's Metaphysics'. *The Review of Metaphysics*, 62(2), 327–54.

Allan, George. 2010. 'In Defense of Secularizing Whitehead'. *Process Studies*, 39(2), 319–33.

Allen, Barry. 2013. 'The Use of Useless Knowledge: Bergson Against the Pragmatists'. *Canadian Journal of Philosophy*, 43(1), 37–59.

Alliez, Eric. 2004. *The Signature of the World, Or, What Is Deleuze and Guattari Philosophy?* Trans. Eliot Ross Albert and Albert Toscano. London and New York: Continuum.

Ansell Pearson, Keith. 1999. *Germinal Life: The Difference and Repetition of Deleuze*. London and New York: Routledge.

Appiah, Kwame Anthony. 2018. *The Lies that Bind: Rethinking Identity*. New York: Liveright Publishing.

Aristotle. 1987. *A New Aristotle Reader*. Ed. J. L. Ackrill. Princeton: Princeton University Press.

Atkins, Brent. 2016. 'Who Thinks Abstractly? Deleuze on Abstraction'. *Journal of Speculative Philosophy*, 30(3), 352–60.

Atleo, E. Richard (Umeek). 2004. *Tsawalk: A Nuu-chah-nulth Worldview*. Vancouver: University of British Columbia Press.

Auxier, Randall E. and Herstein, Gary L. 2017. *The Quantum of Explanation: Whitehead's Radical Empiricism*. New York: Routledge.

Basile, Francesco. 2018. *Whitehead's Metaphysics of Power: Reconstructing Modern Philosophy*. Edinburgh: Edinburgh University Press.

Bateson, Gregory. 2000 [1972]. *Steps to an Ecology of Mind*. Chicago: University of Chicago Press.

Beckett, Samuel. 1989. *Nohow On: Company, Ill Seen Ill Said, Worstward Ho*. New York: Grove Press.

Bell, Jeffrey A. 2009. *Deleuze's Hume: Philosophy, Culture, and the Scottish Enlightenment*. Edinburgh: Edinburgh University Press.

Bell, Jeffrey A. 2014. 'Scientism and the Modern World', in Nicholas Gaskill and A. J. Nocek (eds), *The Lure of Whitehead*. Minneapolis: University of Minnesota Press.

Bennett, Jane. 2010. *Vibrant Matter: A Political Ecology of Things*. Durham, NC: Duke University Press.

Bennett, Jane. 2015. 'Systems and Things: On Vital Materialism and Object-Oriented Philosophy', in Richard Grusin (ed.), *The Nonhuman Turn*. Minneapolis: University of Minnesota Press.

Bergson, Henri. 1935. *The Two Sources of Morality and Religion*. Trans. R. Ashley Audra, Cloudesley Brereton and W. Horsfall Carter. New York: Henry Holt and Company.

Berkeley, George. 2003 [1710]. *A Treatise Concerning the Principles of Human Knowledge*. Mineola, NY: Dover Press.

Berto, Francesco. 2009. *There's Something About Gödel: The Complete Guide to the Incompleteness Theorem*. Malden, MA: Wiley-Blackwell.

Bogue, Ronald. 2007. 'Bergsonian Fabulation and the People to Come', in *Deleuze's Way: Essays in Transverse Ethics and Aesthetics*. Burlington, VT: Ashgate Publishing House.

Bookchin, Murray. 1997. *The Murray Bookchin Reader*. Ed. Janet Biehl. London: Cassell.

Bookchin, Murray. 1998. 'What is Social Ecology?' in M. Zimmerman and J. Baird Callicott (eds), *Environmental Philosophy: From Animal Rights to Radical Ecology* (2nd edn). Upper Saddle River, NJ: Prentice Hall, pp. 462–78.

Bowden, Sean. 2011. *The Priority of Events: Deleuze's Logic of Sense*. Edinburgh: Edinburgh University Press.

Bowden, Sean, Bignall, Simone and Patton, Paul (eds). 2014. *Deleuze and Pragmatism*. London and New York: Routledge.

Braidotti, Rosi. 2011. *Nomadic Theory: The Portable Rosi Braidotti*. New York: Columbia University Press.

Braidotti, Rosi. 2013. *The Posthuman*. Cambridge: Polity.

Braver, Lee. 2007. *A Thing of This World: A History of Continental Anti-Realism*. Evanston, IL: Northwestern University Press.

Bryant, Levi R. 2008. *Difference and Givenness: Deleuze's Transcendental Empiricism and the Ontology of Immanence*. Evanston, IL: Northwestern University Press.

Center for Climate and Energy Solutions. 2019. 'Extreme Weather and Climate Change'. https://www.c2es.org/content/extreme-weather-and-climate-change/ (accessed 15 February 2019).

Climate Communication Science and Outreach. 'Current Extreme Weather and Climate Change'. https://www.climatecommunication.org/new/features/extreme-weather/ overview/ (accessed 15 February 2019).

Code, Lorraine. 2006. *Ecological Thinking: The Politics of Epistemic Location*. New York: Oxford University Press.

Combes, Muriel. 2013 [1999]. *Gilbert Simondon and the Philosophy of the Transindividual*. Trans. Thomas LaMarre. Cambridge, MA: The MIT Press.

Coole, Diana and Frost, Samantha. 2010. *New Materialisms: Ontology, Agency, and Politics*. Durham: Duke University Press.

Cordova, Viola F. 2007. *How It Is: The Native American Philosophy of V. F. Cordova*. Ed. Kathleen Dean Moore, Kurt Peters, Ted Jojola and Amber Lacy. Tucson: University of Arizona Press.

Crutzen, P. 2002. 'Geology of Mankind'. *Nature*, 415(6867), 23.

Culp, Andrew. 2016. *Dark Deleuze* (Forerunners (Minneapolis, MN)). Minneapolis: University of Minnesota Press.

De Freitas, Elizabeth. 2013. 'The Mathematical Event: Mapping the Axiomatic and the Problematic in School Mathematics'. *Studies in Philosophy and Education*, 32(6), 581–99.

De Looper, Christian. 2015. 'Scientists Create Cyborg Roach – Is Cyborg Tech The Way Of The Future?' *Tech Times*. Retrieved from: http://www.techtimes.com/ articles/37719/20150305/scientists-create-cyborg-roach-tech-way-future.htm

Debaise, Didier. 2013. 'A Philosophy of Interstices: Thinking Subjects and Societies from Whitehead's Philosophy'. *Subjectivity*, 6(1), 101–11.

Debaise, Didier. 2017a [2006]. *Speculative Empiricism: Revisiting Whitehead*. Trans. Tomas Weber. Edinburgh: Edinburgh University Press.

Debaise, Didier. 2017b [2015]. *Nature as Event: The Lure of the Possible*. Trans. Michael Halewood. Durham, NC: Duke University Press.

DeLanda, Manuel. 2002. *Intensive Science and Virtual Philosophy*. New York: Bloomsbury Academic.

DeLanda, Manuel and Gaffney, Peter. 2010. 'The Metaphysics of Science: An Interview with Manuel DeLanda', in P. Gaffney (ed.), *The Force of the Virtual: Deleuze, Science, and Philosophy*. Minneapolis: University of Minnesota Press.

Deleuze, Gilles. 1987. *Dialogues*. Trans. and ed. Claire Parnet, Hugh Tomlinson and Barbara Habberjam. New York: Columbia University Press.

Deleuze, Gilles. 1989 [1985]. *The Time-Image: Cinema II*. Trans. Hugh Tomlinson and Robert Galeta. Minneapolis: University of Minnesota Press.

Deleuze, Gilles. 1990 [1969]. *The Logic of Sense*. Trans. Mark Lester. New York: Columbia University Press.

Deleuze, Gilles. 1991 [1953]. *Empiricism and Subjectivity: An Essay on Hume's Theory of Human Nature*. Trans. Constantin Boundas. New York: Columbia University Press.

Deleuze, Gilles. 1993 [1988]. *The Fold: Leibniz and the Baroque*. Trans. Tom Conley. Minneapolis: University of Minnesota Press.

Deleuze, Gilles. 1994 [1969]. *Difference and Repetition*. Trans. Paul Patton. New York: Columbia University Press.

Deleuze, Gilles. 2001. *Pure Immanence: Essays on a Life*. Trans. Anne Boyman. New York; Cambridge, MA: Zone Books; distributed by The MIT Press.

Deleuze, Gilles. 2004. *Desert Islands*. Ed. David Lapoujade. Trans. Michael Taormina. Los Angeles: Semiotext(e).

Deleuze, Gilles and Guattari, Félix. 1983 [1972]. *Anti-Oedipus: Capitalism and Schizophrenia*. Trans. Robert Hurley, Mark Seem and Helen R. Lane. Minneapolis: University of Minnesota Press.

Deleuze, Gilles and Guattari, Félix. 1987 [1980]. *A Thousand Plateaus: Capitalism and Schizophrenia*. Trans. Brian Massumi. Minneapolis: University of Minnesota Press.

Deleuze, Gilles and Guattari, Félix. 1994 [1991]. *What Is Philosophy?* Trans. Hugh Tomlinson and Graham Burchell. New York: Columbia University Press.

Derrida, Jacques. 1981. *Dissemination*. Chicago: University of Chicago Press.

Descola, Philippe. 2013 [2005]. *Beyond Nature and Culture*. Trans. Janet Lloyd. Chicago and London: The University of Chicago Press.

Dewey, John. 1937. 'Whitehead's Philosophy'. *The Philosophical Review*, 46(2), 170–7.

Dombrowski, Daniel. A. 2017. *Whitehead's Religious Thought: From Mechanism to Organism, From Force to Persuasion*. Albany: SUNY Press.

Duvernoy, Russell J. 2019. 'Deleuze, Whitehead, and the "Beautiful Soul"'. *Deleuze and Guattari Studies*, 13(2), 163–85.

Duvernoy, Russell J. 2020. 'Feeling Facts with Whitehead and Deleuze', in *Process Thought* (forthcoming). Berlin: Ontos Verlag.

Evens, Aden. 2000. 'Math Anxiety'. *Angelaki*, 5(3), 105–15.

Faber, Roland, Bell, Jeffrey and Petek, Joseph. 2017. *Rethinking Whitehead's Symbolism: Thought, Language, Culture*. Edinburgh: Edinburgh University Press.

Fischetti, Mark. 2015. 'Climate Change Hastened Syria's Civil War'. *Scientific American*. Retrieved from: https://www.scientificamerican.com/article/climate-change-hastened-the-syrian-war/

Foer, Franklin. 2017. *World Without Mind: The Existential Threat of Big Tech*. New York: Penguin Books.

Foucault, Michel. 1994 [1966]. *The Order of Things: An Archaeology of the Human Sciences*. New York: Vintage Books.

Francis, Pope. 2015. *Encyclical by Pope Francis: Laudato Si' (Be Praised): On Care for Our Common Home*. Rome: Council on Foreign Relations.

Gaskill, Nicholas and Nocek, A. J. 2014. *The Lure of Whitehead*. Minneapolis: University of Minnesota Press.

Gelman, Susan. 2004. 'Psychological Essentialism in Children'. *Trends in Cognitive Sciences*, 8(9), 404–9.

Ghiselin, Michael T. 1997. *Metaphysics and the Origin of Species*. Albany: SUNY Press.

Ghosh, Amitav. 2016. *The Great Derangement: Climate Change and the Unthinkable*. Chicago: The University of Chicago Press.

Gleick, Peter H. 2014. 'Water, Drought, Climate Change and Conflict in Syria'. *Weather, Climate and Society*, 6(3). Retrieved from: http://journals.ametsoc.org/doi/full/10.1175/WCAS-D-13-00059.1

Gracia, Jorge J. E. 1988. *Individuality: An Essay on the Foundations of Metaphysics*. Albany: SUNY Press.

Griffin, David Ray. 2016. *Protecting Our Common, Sacred Home: Pope Francis and Process Thought*. Anoka, MN: Process Century Press.

Grosz, E. A. 2008. *Chaos, Territory, Art: Deleuze and the Framing of the Earth*. New York: Columbia University Press.

Grusin, Richard (ed.). 2015. *The Nonhuman Turn*. Minneapolis: University of Minnesota Press.

Guattari, Félix. 1995. *Chaosmosis: An Ethico-aesthetic Paradigm*. Bloomington: Indiana University Press.

Guattari, Félix. 2008 [1989]. *The Three Ecologies*. Trans. Ian Pindar and Paul Sutton. New York: Continuum Press.

Hallward, Peter. 2006. *Out of This World: Deleuze and the Philosophy of Creation*. London: Verso.

Hansen, Mark B. N. 2015a. *Feed-Forward: On the Future of Twenty-First Century Media*. Chicago: University of Chicago Press.

Hansen, Mark B. N. 2015b. 'Our Predictive Condition; or, Prediction in the Wild', in Richard Grusin (ed.), *The Nonhuman Turn*. Minneapolis: University of Minnesota Press.

Haraway, Donna. 2001 [1984]. 'A Cyborg Manifesto', in David Bell and Barbara Kennedy (eds), *The Cybercultures Reader*. New York: Routledge.

Harman, Graham. 2014a. 'Whitehead and Schools X, Y, and Z', in Nicholas Gaskill and A. J. Nocek (eds), *The Lure of Whitehead*. Minneapolis: University of Minnesota Press.

Harman, Graham. 2014b. 'Another Response to Shaviro', in Roland Faber and Andrew Goffey (eds), *The Allure of Things: Process and Object in Contemporary Philosophy*. London and New York: Bloomsbury Press.

Harney, Stefano and Moten, Fred. 2013. *The Undercommons: Fugitive Planning and Black Study*. Brooklyn: Minor Compositions.

Hartshorne, Charles. 1973. *Born to Sing: An Interpretation and World Survey of Bird Song*. Bloomington: Indiana University Press.

Hartshorne, Charles. 1997. *The Zero Fallacy and Other Essays in Neoclassical Philosophy*. Chicago, IL: Open Court Publishing.

Haugeland, John. 1982. 'Heidegger on Being a Person'. *Noûs*, 16(1), 15–26.

Hazlett, Allan. 2010. 'Brutal Individuation', in Allan Hazlett (ed.), *New Waves in Metaphysics*. London: Palgrave Macmillan.

Hegel, G. W. F. 1977 [1807]. *Phenomenology of Spirit*. Trans. A. V. Miller. Oxford: Oxford University Press.

Heidegger, Martin. 1995 [1983]. *The Fundamental Concepts of Metaphysics: World, Finitude, Solitude*. Trans. William McNeill and Nicholas Walker. Bloomington: Indiana University Press.

Heidegger, Martin. 2008 [1927]. *Being and Time*. Trans. John Macquarrie and Edward Robinson. New York: HarperCollins.

Henning, Brian G. 2005. *The Ethics of Creativity: Beauty, Morality, and Nature in a Processive Cosmos*. Pittsburgh: University of Pittsburgh Press.

Hocking, W. E. 1963. 'Whitehead as I Knew Him', in G. K. Kline (ed.), *Alfred North Whitehead: Essays on his Philosophy*. Englewood Cliffs, NJ: Prentice Hall.

Husserl, Edmund. 1931. *Ideas: General Introduction to a Pure Phenomenology*. Trans. W. R. Boyce Gibson. New York: Macmillan.

Intergovernmental Panel on Climate Change (IPCC). 2018. *Global Warming of 1.5°C: An IPCC Special Report on the Impacts of Global Warming of 1.5°C above Pre-Industrial Levels and Related Global Greenhouse Gas Emission Pathways, in the Context of Strengthening the Global Response to the Threat of Climate Change, Sustainable Development, and Efforts to Eradicate Poverty.* Geneva: World Meteorological Organization. Available at: https://report.ipcc.ch/sr15/pdf/sr15_spm_final.pdf

James, William. 1976 [1912]. *Essays in Radical Empiricism.* Cambridge, MA: Harvard University Press.

James, William. 1977 [1908]. *A Pluralistic Universe.* Cambridge, MA: Harvard University Press.

James, William. 1988. *Manuscript Essays and Notes.* Cambridge, MA: Harvard University Press.

Jardine, Alice. 1984. 'Woman in Limbo: Deleuze and His Br(others)'. *SubStance, 13*(3/4, issue 44–5), 46–60.

Jones, Jude. 1998. *Intensity: An Essay in Whiteheadian Ontology.* Nashville: Vanderbilt University Press.

Kant, Immanuel. 1987 [1790]. *Critique of Judgement.* Trans. Werner S. Pluhar. Indianapolis: Hackett Publishing.

Kant, Immanuel. 1997 [1783]. *Prolegomena to Any Future Metaphysics.* Ed. Gary Hatfield. Cambridge: Cambridge University Press.

Kant, Immanuel. 1998 [1787]. *Critique of Pure Reason.* Trans. Paul Guyer and Allen Wood. Cambridge: Cambridge University Press.

Kerslake, Christian. 2009. *Immanence and the Vertigo of Philosophy: From Kant to Deleuze.* Edinburgh: Edinburgh University Press.

Klein, Alexander. 2009. 'On Hume on Space: Green's Attack, James' Empirical Response'. *Journal of the History of Philosophy, 47*(3), 415–49.

Kline, George L. 1983. 'Form, Concrescence, Concretum', in Lewis S. Ford and George L. Kline (eds), *Explorations in Whitehead's Philosophy.* New York: Fordham University Press.

Kohn, Eduardo. 2013. *How Forests Think: Toward an Anthropology Beyond the Human.* Berkeley: University of California Press.

Krause, Bernie. 2009. 'The Great Animal Orchestra'. Lecture at California Academy of Sciences. Retrieved from: http://library.fora.tv/2009/09/22/Dr_Bernie_Krause_The_Great_Animal_Orchestra

Krauss, Elizabeth. 1998. *The Metaphysics of Experience: A Companion to Whitehead's Process and Reality.* New York: Fordham University Press.

Kripke, Saul. 1980. *Naming and Necessity.* Cambridge, MA: Harvard University Press.

Kuperus, Gerard. 2007. 'Heidegger and Animality', in Christian Lotz and Corinne Painter (eds), *Phenomenology and the Non-Human Animal.* New York: Kluwer/Springer.

Lamberth, David C. 1999. *William James and the Metaphysics of Experience.* Cambridge: Cambridge University Press.

Lapoujade, David. 2000. 'From Transcendental Empiricism to Worker Nomadism: William James'. *Pli*, 9, 190–9.

Latour, Bruno. 1991. *We Have Never Been Modern.* Trans. Catherine Porter. Cambridge, MA: Harvard University Press.

Latour, Bruno. 2017 [2015]. *Facing Gaia: Eight Lectures on the New Climatic Regime.* Trans. Catherine Porter. Medford, MA: Polity Press.

Le Doeuff, Michèle. 1989 [1980]. *The Philosophical Imaginary.* Trans. Colin Gordon. Stanford: Stanford University Press.

Lear, Jonathan. 2000. *Happiness, Death, and the Remainder of Life.* Cambridge, MA: Harvard University Press.

Lewontin, Richard. 2000. *The Triple Helix: Gene, Organism, and Environment.* Cambridge, MA: Harvard University Press.

Lewontin, Richard, Rose, Steven and Kamin, Leon. 2000 [1984]. 'Not in Our Genes', in L. Stevenson (ed.), *The Study of Human Nature: A Reader*. New York: Oxford University Press.

Livingston, Paul. 2012. *The Politics of Logic: Badiou, Wittgenstein, and the Consequences of Formalism*. Routledge Studies in Contemporary Philosophy, 27. New York: Routledge.

Locke, John. 2008 [1687]. *An Essay Concerning Human Understanding*. Oxford: Oxford University Press.

Lovejoy, Arthur O. 1936. *The Great Chain of Being: A Study of the History of an Idea*. Cambridge, MA: Harvard University Press.

Lowe, E. J. 2003. 'Individuation', in Michael J. Loux and Dean W. Zimmerman (eds), *The Oxford Handbook of Metaphysics*. Oxford: Oxford University Press.

McHenry, Leemon B. 2015. *The Event Universe: The Revisionary Metaphysics of Alfred North Whitehead*. Edinburgh: Edinburgh University Press.

Madelrieux, Stephane. 2014. 'Pluralism without Pragmatism: Deleuze and the Ambiguities of the French Reception of James', in Sean Bowden, Simone Bignall and Paul Patton (eds), *Deleuze and Pragmatism*. London and New York: Routledge.

Malabou, Catherine. 2008 [2004]. *What Should We Do with Our Brain?* Trans. Sebastian Rand. New York: Fordham University Press.

Manning, Erin. 2013. *Always More than One: Individuation's Dance*. Durham, NC: Duke University Press.

Manning, Erin. 2015. 'Artfulness', in Richard Grusin (ed.), *The Nonhuman Turn*. Minneapolis: University of Minnesota Press.

Marx, Karl. 1992 [1867]. *Capital: A Critique of Political Economy*. Vol. 1. Trans. Ben Fowkes. New York: Penguin.

Marx, Karl. 1994. *Selected Writings*. Ed. Lawrence H. Simon. Indianapolis: Hackett Publishing Company.

Massumi, Brian. 2002. *Parables for the Virtual: Movement, Affect, Sensation* (Post-Contemporary Interventions). Durham, NC: Duke University Press.

Massumi, Brian. 2014. *What Animals Teach Us about Politics*. Durham, NC: Duke University Press.

Massumi, Brian. 2015. 'The Supernormal Animal', in Richard Grusin (ed.), *The Nonhuman Turn*. Minneapolis: University of Minnesota Press.

Massumi, Brian. 2018. *99 Theses on the Revaluation of Value: A Postcapitalist Manifesto*. Minneapolis: University of Minnesota Press.

Mautner, Thomas. 2005. *The Penguin Dictionary of Philosophy* (2nd edn). London; New York: Penguin.

Mayr, Ernst. 1961. 'Cause and Effect in Biology'. *Science*, 134(3489), 1501–6.

Meillassoux, Quentin. 2008 [2006]. *After Finitude: An Essay on the Necessity of Contingency*. Trans. Ray Brassier. London and New York: Bloomsbury Academic Press.

Misak, C. J. 2013. *The American Pragmatists*. Oxford: Oxford University Press.

Morar, Nicolae, Toadvine, Ted and Bohannan, Brendan J. M. 2015. 'Biodiversity at Twenty-Five Years: Revolution Or Red Herring?' *Ethics, Policy & Environment*, 18(1), 16–29. DOI: 10.1080/21550085.2015.1018380

Naess, Arne and Rothenberg, David. 1989. *Ecology, Community, and Lifestyle: Outline of an Ecosophy*. Cambridge and New York: Cambridge University Press.

Nancy, Jean-Luc. 2015. *After Fukushima: The Equivalence of Catastrophes*. New York: Fordham University Press.

Nature Editorial Board. 2018, 13 February. 'Don't Jump to Conclusions about Climate Change and Civil Conflict'. *Nature*, 554, 275–6. DOI: 10.1038/d41586-018-01875-9. Retrieved from: https://www.nature.com/articles/d41586-018-01875-9 (18 June 2020).

Nobo, Jorge Luis. 1986. *Whitehead's Metaphysics of Extension and Solidarity*. Albany: SUNY Press.

Norton-Smith, Thomas. 2010. *The Dance of Person and Place: One Interpretation of American Indian Philosophy*. SUNY Series in Living Indigenous Philosophies. Albany: SUNY Press.

O'Hara, Frank. 1976. *The Selected Poems of Frank O'Hara*. New York: Vintage Books.

Oury, Jean. 1982. *L'Aliénation*. Paris: Galilée.

Oury, Jean. 1989. *Création et schizophrénie*. Paris: Galilée.

Oury, Nouvelle and Reggio, David. 2007. 'The Hospital is Ill'. *Radical Philosophy*, *143*, 32–45.

Pojman, Louis and Pojman, Paul (eds). 2012. *Environmental Ethics: Readings in Theory and Application* (6th edn). Belmont, CA: Wadsworth Publishing.

Priest, Graham. 1995. *Beyond the Limits of Thought*. Cambridge: Cambridge University Press. (2nd expanded edition, Oxford: Oxford University Press, 2002.)

Priest, Graham. 2008. *An Introduction to Non-classical Logic: From If to Is* (2nd edn, Cambridge Introductions to Philosophy). Cambridge and New York: Cambridge University Press.

Proctor, R. and Schiebinger, Londa L. 2008. *Agnotology: The Making and Unmaking of Ignorance*. Stanford: Stanford University Press.

Protevi, John. 2009. *Political Affect: Connecting the Social and the Somatic*. Minneapolis: University of Minnesota Press.

Raison, C., Lowry, C. and Rook, G. 2010. 'Inflammation, Sanitation, and Consternation Loss of Contact With Coevolved, Tolerogenic Microorganisms and the Pathophysiology and Treatment of Major Depression'. *Archives Of General Psychiatry*, *67*(12), 1211–24.

Rajamannar, 2017. 'How to Create a Campaign that Spans 2 Decades'. Adweek.com. Available at: https://www.adweek.com/brand-marketing/how-to-create-a-campaign-that-spans-2-decades-like-mastercards-priceless/ (last accessed 21 November 2019).

Rescher, Nicholas. 1996. *Process Metaphysics: An Introduction to Process Philosophy*. Albany: SUNY Press.

Robinson, Keith. 2009. *Deleuze, Whitehead, Bergson: Rhizomatic Connections*. Basingstoke: Palgrave Macmillan.

Rorty, Richard. 1979. *Philosophy and the Mirror of Nature*. Princeton: Princeton University Press.

Rorty, Richard. 1983. 'Unsoundness in Perspective', *Times Literary Supplement*, *17*, 619–20.

Roth, E. and Pouty, B. 1985. *Méthodes de datation par les phénomènes nucléaires naturels*. Paris: Masson.

Russell, Bertrand. 1961. *History of Western Philosophy*. London: George Allen and Unwin Ltd.

Ruyer, Raymond. 2016 [1952]. *Neofinalism*. Trans. Alyosha Edlebi. Minneapolis: University of Minnesota Press.

Sarkar, Sahotra. 2005. 'Ecology', in Stanford Encyclopedia of Philosophy. Available at: https://plato.stanford.edu/entries/ecology/ (last accessed 21 November 2019).

Sassen, Saskia. 2014. *Expulsions: Brutality and Complexity in the Global Economy*. Cambridge, MA: The Belknap Press of Harvard University Press.

Scranton, Roy. 2015. *Learning to Die in the Anthropocene*. San Francisco: City Lights Books.

Sedgwick, Eve Kosofsky, Frank, Adam and Alexander, Irving E. 1995. *Shame and Its Sisters: A Silvan Tomkins Reader*. Durham, NC: Duke University Press.

Seigfried, Charlene Haddock. 1990. *William James's Radical Reconstruction of Philosophy*. Albany: SUNY Press.

Serres, Michel. 1995 [1990]. *The Natural Contract*. Trans. Elizabeth MacArthur and William Paulson. Ann Arbor: University of Michigan Press.

Shaviro, Steven. 2009. *Without Criteria: Kant, Whitehead, Deleuze, and Aesthetics*. Technologies of Lived Abstraction. Cambridge, MA: The MIT Press.

Shaviro, Steven. 2014. *The Universe of Things: On Speculative Realism*. Minneapolis: University of Minnesota Press.

Shaviro, Steven. 2015. 'Consequences of Panpsychism', in Richard Grusin (ed.), *The Nonhuman Turn*. Minneapolis: University of Minnesota Press.

Simondon, Gilbert. 1989. *L'individuation psychique et collective: À la lumière des notions de forme, information, potentiel et métastabilite*. Paris: L'invention philosophique, Aubier.

Smith, Daniel W. 2012. *Essays on Deleuze*. Edinburgh: Edinburgh University Press.

Snyder, Gary. 1990. *The Practice of the Wild*. Berkeley: Counterpoint Press.

Spinoza, Baruch. 1992. *The Ethics; Treatise on the Emendation of the Intellect; Selected Letters*. Ed. Samuel Shirley and Seymour Feldman. Indianapolis: Hackett Publishing Co.

Stauffer, Robert C. 1957. 'Haeckel, Darwin, and Ecology'. *The Quarterly Review of Biology, 32*(2), 138–44.

Steffen, W., Grinevald, J., Crutzen, P. and McNeill, J. 2011. 'The Anthropocene: Conceptual and Historical Perspectives'. *Philosophical Transactions of The Royal Society A – Mathematical Physical And Engineering Sciences, 369*(1938), 842–67.

Stengers, Isabelle. 2010 [2003]. *Cosmopolitics I*. Minneapolis: University of Minnesota Press.

Stengers, Isabelle. 2011 [2002]. *Thinking with Whitehead: A Free and Wild Creation of Concepts*. Cambridge, MA: Harvard University Press.

Stengers, Isabelle. 2014. 'Speculative Philosophy and the Art of Dramatization', in Roland Faber and Andrew Goffey (eds), *The Allure of Things: Process and Object in Contemporary Philosophy*. London and New York: Bloomsbury Press.

Stengers, Isabelle. 2015 [2009]. *In Catastrophic Times: Resisting the Coming Barbarism*. Trans. Andrew Goffey. London: Open Humanities Press.

Stern, Daniel. 1985. *The Interpersonal World of the Infant*. New York: Basic Books.

Stern, Daniel. 2010. *Forms of Vitality: Exploring Dynamic Experience in Psychology, the Arts, Psychotherapy, and Development*. Oxford and New York: Oxford University Press.

Stiegler, Bernard. 2014. 'On a Positive Pharmacology'. *Rue Descartes, 82*(3), 1325.

Strawson, P. 1959. *Individuals: An Essay in Descriptive Metaphysics*. London: Methuen.

Sullivan, Shannon and Tuana, Nancy. 2007. *Race and Epistemologies of Ignorance*. Albany: SUNY Press.

Thomson, Iain. 2004. 'Ontology and Ethics at the Intersection of Phenomenology and Environmental Philosophy'. *Inquiry, 47*(4), 380–412.

Toscano, Alberto. 2006. *The Theatre of Production: Philosophy and Individuation between Kant and Deleuze*. New York: Palgrave Macmillan.

Uexküll, Jakob von. 2010 [1934]. *A Foray into the Worlds of Animals and Humans*. Trans. Joseph D. O'Neil. Minneapolis: University of Minnesota Press.

United Nations Refugee Agency. 'Climate Change and Disasters'. Available at: https://www.unhcr.org/climate-change-and-disasters.html (accessed 15 February 2019).

United Nations Refugee Agency. 'Figures at a Glance'. Available at: https://www.unhcr.org/figures-at-a-glance.html (accessed 15 February 2019).

Van Valen, L. 1973. 'Festschrift'. *Science, 180*, 488.

Wahl, Jean. 1925 [1920]. *The Pluralist Philosophies of England & America*. Trans. Fred Rothwell. London: The Open Court Company.

Wahl, Jean. 2004 [1932]. *Vers le Concret*. Deuxième édition augmentée. Paris: Librairie Philosophique J. Vrin, Bibliothèque d'Histoire de la Philosophie.

Wahl, Jean. 2016. *Human Existence and Transcendence*. Trans. William C. Hackett. Notre Dame, IN: University of Notre Dame Press.

Wahl, Jean. 2017. *Transcendence and the Concrete: Selected Writings*. Ed. Alan D. Schrift and Ian Alexander Moore. New York: Fordham University Press.

Washington University in St. Louis. 2016, June 28. 'Engineers to Use Cyborg Insects as Biorobotic Sensing Machines'. *ScienceDaily*. Retrieved from: www.sciencedaily.com/releases/2016/06/160628141415.htm (17 February 2017).

Whitehead, Alfred North. 1920. *The Concept of Nature*. Cambridge: Cambridge University Press.

Whitehead, Alfred North. 1958 [1929]. *The Function of Reason*. Boston, MA: Beacon Press.

Whitehead, Alfred North. 1966 [1938]. *Modes of Thought*. New York: Free Press.

Whitehead, Alfred North. 1967 [1925]. *Science and the Modern World*. New York: Free Press.

Whitehead, Alfred North. 1967 [1933]. *Adventures of Ideas*. New York: Free Press.

Whitehead, Alfred North. 1978 [1929]. *Process and Reality: An Essay in Cosmology* (Corrected edn). New York: Free Press.

Whitehead, Alfred North. 1985 [1927]. *Symbolism: Its Meaning and Effect*. New York: Fordham University Press.

Willett, Cynthia. 2014. *Interspecies Ethics*. New York: Columbia University Press.

Williams, James. 2005. *The Transversal Thought of Gilles Deleuze: Encounters and Influences*. Manchester: Clinamen Press.

Williams, James. 2006. 'Science and Dialectics in the Philosophies of Deleuze, Bachelard and DeLanda'. *Paragraph*, *29*(2), 98–114.

Williams, James. 2008. *Gilles Deleuze's Logic of Sense: A Critical Introduction and Guide*. Edinburgh: Edinburgh University Press.

Williams, James. 2009. 'A. N. Whitehead', in Graham Jones and Jon Roffe (eds), *Deleuze's Philosophical Lineage*. Edinburgh: Edinburgh University Press.

Williams, James. 2010. 'Against Oblivion and Simple Empiricism: Gilles Deleuze's "Immanence: A Life . . ."'. *Journal of Philosophy: A Cross-Disciplinary Inquiry*, *5*(11), 25–34.

Worster, Donald. 1977. *Nature's Economy: A History of Ecological Ideas*. New York: Cambridge University Press.

Young, Robert M. 1968. 'The Association of Ideas', in Philip P. Weiner (ed.), *Dictionary of the History of Ideas*. New York: Scribners.

Zamberlin, Mary F. 2006. *Rhizosphere: Gilles Deleuze and the 'Minor' American Writings of William James, W.E.B. Du Bois, Gertrude Stein, Jean Toomer, and William Faulkner*. New York: Routledge.

Žižek, Slavoj. 2004. *Organs without Bodies: Deleuze and Consequences*. London: Routledge.

Zourabichvili, François. 2012. *Deleuze, a Philosophy of the Event: Together with the Vocabulary of Deleuze*. Trans. Kieran Aarons. Edinburgh: Edinburgh University Press.

Index